Synthesis Lectures on Engineering, Science, and Technology

The focus of this series is general topics, and applications about, and for, engineers and scientists on a wide array of applications, methods and advances. Most titles cover subjects such as professional development, education, and study skills, as well as basic introductory undergraduate material and other topics appropriate for a broader and less technical audience.

Ricardo Madeira · João Pedro Oliveira ·
Nuno Paulino

Fully Integrated Switched-Capacitor PMU for IoT Nodes

Analysis and Design

 Springer

Ricardo Madeira (iD)
CTS-UNINOVA/FCT-NOVA
Caparica, Portugal

João Pedro Oliveira
CTS-UNINOVA/FCT-NOVA
Caparica, Portugal

Nuno Paulino (iD)
CTS-UNINOVA/FCT-NOVA
Caparica, Portugal

ISSN 2690-0300   ISSN 2690-0327 (electronic)
Synthesis Lectures on Engineering, Science, and Technology
ISBN 978-3-031-14703-6   ISBN 978-3-031-14701-2 (eBook)
https://doi.org/10.1007/978-3-031-14701-2

This Springer imprint is published by the registered company Springer Nature Switzerland AG
The registered company address is: Gewerbestrasse 11, 6330 Cham, Switzerland

Preface

The IoT smart nodes deployment has been increasing rapidly in recent years, in a wide range of applications. These nodes are typically inserted in WSN, ranging from a few numbers to hundreds of nodes. Solutions using SoC have been proposed for the node implementation, due to their low production costs. These SoC are typically composed of sensors, to collect the data from the real world; an analogue front-end system for signal processing (S&H, Filter, and ADC/DAC) to convert the collected data to the digital domain; a microprocessor unit to do some signal processing of the data; a communication unit to transmit the treated data; and finally, a PMU, which is the focus of this book, to efficiently power all these blocks. Since the available energy is limited, efficient energy conversion is crucial to reduce maintenance costs. To this end, SC DC-DC converters have been proposed, since they can be fully integrated into CMOS technology, and offer a good trade-off between efficiency and power density. This book describes the design and test of a fully integrated PMU IC prototype, implemented in 130 nm bulk CMOS technology. The prototype is a fully integrated 16 mW PMU which includes a multi-ratio 1+3 binary-weighted SC converter, composed of three voltage CR 1/2, 2/3, and 1/1, covering an input voltage range of 1.1 V–2.3 V, and generating a 0.9 V output voltage. Time interleaved and capacitance modulation techniques were employed to completely remove the external decoupling capacitor. The output voltage ripple is further reduced by adjusting the number of active cells, according to the output power level and input voltage value, sensed through the average clock frequency. The PMU includes a set of auxiliary circuits to sustain the converter operation. These are a phase generator, a CR controller, the switch drivers, a cell controller, a voltage reference generator, and a start-up circuit. The total circuit active area is 5.12 mm^2 and a peak efficiency of 74.3% was measured.

Lisbon, Portugal

Ricardo Madeira

Acknowledgements

The authors would like to acknowledge many people who directly, or indirectly, supported and contributed to the development of this work. We specially thank to João Goes, Luis Olivera, Jorge Fernandes, Rui Tavares, Hugo Serra, Nuno Correia, Nuno Pereira, Miguel Fernandes, Miguel Teixeira, Ana Correia, Błazej Nowacki, and João Melo.

This work was supported by the Fundação para a Ciência e Tecnologia of Ministério da Ciência e Tecnologia e Ensino Superior, through the Ph.D. grant SFRH/BD/115543/2016. Also by the following projects: PROTEUS funded by the European Union's 2020 Programme under Grant 644852, CTS multi-annual funding program funds project PEST (PEST-OE/EEI/UI0066/2013, and PEST-OE/EEI/UI0066/2014), foRESTER PCIF/SSI/0102/2017, and UIDB/00066/2020.

Contents

Acronyms

AC	Auxiliary SC cell
APS	Spectre Accelerated Parallel Simulator
ASM	Asynchronous state machine
ASP	Algebraic Series-Parallel
AVFI	Algorithmic Voltage-Feed-In
BOM	Bill of Materials
CMOD	Carrier Module
CMOS	Complementary Metal-Oxide-Semiconductor
CR	Conversion Ratio
CRS	Charge Redistribution Step
CTAT	Complementary to Absolute Temperature
EH	Energy Harvesting
ESD	Electrostatic discharge
ff	Fast-Fast
fnsp	Fast NMOS-Slow PMOS
FPGA	Field Programmable Gate Array
GBW	Gain-Bandwidth
IC	Integrated Circuit
IoT	Internet of Things
LDO	Low Drop-Out Regulators
MC	Main SC cell
MIM	Metal-Insulator-Metal
MOM	Metal-Oxide-Metal
MPPT	Maximum Power Point Tracking
MSC	Multiphase Soft-Charging
MUX	Multiplexer
NCG	Non-overlap Clock Generator
NSC	Negator-based SC
PCB	Printed Circuit Board
PDA	Dynamic Power Allocation

PG	Phase Generator
PMU	Power Management Unit
PSRR	Power Supply Rejection Ratio
PTAT	Proportional to Absolute Temperature
PV	Photovoltaic
PVT	Process, Voltage, and Temperature
QF	Charge-path folding
RC	Rational cell
RF	Radio Frequency
RSC	Recursive SC
SAR	Successive-Approximation-Register
SC	Switched Capacitor
SCW	Symmetric Cockcroft-Walton
snfp	Slow NMOS-Fast PMOS
SO	Stage Outphasing
SoA	State-of-the-Art
SoC	System-on-Chip
SOI	Silicon-On-Insulator
SP	Series-Parallel
SPCR	Scalable Parasitic Charge Redistribution
SR	Set-Reset
ss	Slow-Slow
tt	Typical-Typical
VFI	Voltage-feed-in
WSN	Wireless Sensor Network

Introduction

1

1.1 Motivation

The Internet of Things (IoT) plays an important role on the new generation of information technology, with a major impact in the human life in the recent and near future. This still recent technology aims to enable things to be connected anytime, anywhere, with anything and anyone, using any path/network or service [1]. This means that the objects in our environment would become smart, in the sense that they would be able to sense, interpret, share information, and react to events and human activities in the surrounding environment [2]. These smart objects, or smart nodes, are typically connected through a Wireless Sensor Network (WSN), where the node's number can vary from a few nodes, to hundred, or even thousand of nodes [3–5]. The production and maintenance costs must be as low as possible, for such large networks to be implemented. Hence, these nodes must be as autonomous and integrated as possible, and thus System-on-Chip (SoC) solutions have been proposed for their implementation [6–8]. These SoC are typically composed of sensors, to collect the data from the real world; an analogue front-end system for signal processing (e.g. S&H, Filters, ADC/DAC) to convert the collected data to the digital domain; a microprocessor unit to do some signal processing of the data; a communication unit to transmit the treated data; and finally, a Power Management Unit (PMU), which is the focus of this book, to efficiently power all these blocks.

As stated above, the PMU is responsible for powering the smart node. The node can be powered either by using a pre-charged energy storage device, like a battery or a supercapacitor, or by harvesting energy, Energy Harvesting (EH), from the surrounding environment, e.g. Photovoltaic (PV), piezoelectric, thermoelectric, or Radio Frequency (RF) energy [6–11]. Typically, a combination of both the EH and the energy storage device is used, where the harvested energy is stored in the energy stored device, to be then extracted by the PMU, for supplying energy to the system [6, 7]. However, generally the amount of available energy

© The Author(s), under exclusive license to Springer Nature Switzerland AG 2022 1
R. Madeira et al., *Fully Integrated Switched-Capacitor PMU for IoT Nodes*, Synthesis Lectures on Engineering, Science, and Technology,
https://doi.org/10.1007/978-3-031-14701-2_1

from the surrounding environment is reduced, typically only providing a fraction of the power needed for the system operation. Thus, these systems can spend a long time collecting energy and are only powered during small periods of time. For this type of operation, the energy storage device must withstand a large number of charge/discharge cycles without suffering from degradation. Hence, supercapacitors are an interesting solution for IoT nodes [10, 12, 13] because they have, ideally, a limitless number of charge/discharge cycles, and do not require a complex charging circuit, unlike batteries. Nonetheless, a rechargeable battery has the advantage of having a larger energy storing capacity, however, its lifetime is seriously reduced by the number of charge/discharge cycles (typically around 1000 cycles) [13–15]. Considering that typically the amount of available energy is limited, its extraction, either from the surrounding environment or from an energy stored device, must be made as efficient as possible. To this end, PMU using Maximum Power Point Tracking (MPPT) have been developed to efficiently extract the energy from the surrounding environment [6–8, 10, 11, 16, 17], and also PMU with high efficient voltage regulators have been developed to convert a variable voltage from an energy storing device to a constant supply voltage, suitable for the system operation [13, 18–20].

A constant DC voltage can be generated using a linear voltage converter, e.g. a Low Drop-Out Regulators (LDO), or by using a switched-mode DC-DC converter. The LDO have the advantage of being easily fully integrated and achieving high power densities [21, 22]. However, they are unable to step-up the supply voltage and only reach high energy efficiency values when their input voltage is close to its output voltage. As for the switched-mode DC-DC converters, they can be implemented using inductors—inductive converters [23–27]—or using capacitors—Switched Capacitor (SC) converters [19, 20, 28–32]. Both can step-up, or step-down, the voltage supply and can achieve high efficiency when using external inductors or capacitors. However, when fully integrating these converters, the inductive converters have their voltage conversion efficiency and power density reduced, due to the low quality factor and large area of the on-chip inductors [24, 27, 33]. On the other hand, capacitive converters can easily be fully integrated because they are only composed by transistors and capacitors, which are native in bulk Complementary Metal-Oxide-Semiconductor (CMOS) technology. In bulk CMOS technology, these converters achieve medium power density values, on the order of some milliwatts to hundred of milliwatts, and can achieve energy conversion efficiency values between 70 and 90% [19, 20, 28, 30–32]. The fully integration of these converters reduces the overall size of the system, since the discrete bulky capacitors and inductors are removed, thus reducing the Printed Circuit Board (PCB) space occupied by the PMU. This means that the Bill of Materials (BOM) is also reduced, and thus the fabrication and assembly costs are decreased. Furthermore, now the PMU is now much closer to its load. This proximity leads to better voltage regulation and better transient response. Considering this, the SC converters have attracted a significant interest in both the academia and the industry [34].

The working principle of the SC converters is easy to understand, a capacitor is used to transfer charge from the input to the output, where the frequency at which the charge is transferred, defines the converter's output voltage value. Typically, this charge transferring is controlled by two clock phase signals that allow for different circuit configurations on each clock phase. The SC converter has a finite output impedance, and thus the maximum power density is determined by the on-chip capacitance density. Due to the low capacitance density of on-chip bulk CMOS capacitors, there is a trade-off between the energy efficiency and power density [34, 35]. Furthermore, in fully integrated SC converters, minimizing the output voltage ripple comes at the cost of increasing the converter's area or circuit complexity, by using a large decoupling capacitor, or by using multiple cells in a time interleaving scheme, or by using other charge modulation techniques. This increases the design complexity of an SC converter making it a challenging task.

Finally, for an SC converter to operate properly, a set of auxiliary circuits are required. Typically, extra circuits like comparators, reference voltage generators, oscillators, digital circuits, and start-up circuits, are required. These must all be integrated together with the converter which further increases the design complexity of the PMU.

1.2   Outline

This book is composed of 7 chapters, including this introduction. A short description can be found below. In Chap. 2, an overview of the SC DC-DC converter fundamentals is presented. It starts with a detailed description of the passive implementation in 130 nm CMOS technology. Next, a theoretical step-by-step analysis used to size an SC DC-DC converter and determine its performance is depicted. The chapter ends with a brief description of the commonly used loop regulation techniques.

Chapters 3 and 4 cover the State-of-the-Art (SoA) of fully integrated SC DC-DC converters in CMOS technology. The first describes and analyses the performance of the most used topologies in multi-ratio SC converters. The second describes the techniques used to enhance the performance of fully integrated SC converters.

Chapter 5 shows the design process of a fully integrated PMU designed for a maximum output power of 16 mW, capable of converting a supercapacitor's variable input voltage from 1.1 to 2.3 V, whilst generating an output voltage of 0.9 V. The PMU includes a SC DC-DC converter that is dived into $1 + 3$ binary-weighted cells, which are further divided into 32 smaller cells, where time interleaved was applied, to eliminate the external decoupling capacitor. The 3 binary-weighted cells can be enabled or disabled according to the power required at the output and the input voltage, sensed through the clock frequency. The PMU also includes a set of auxiliary circuits to ensure the correct behaviour of the converter, according to the input voltage and the output load. These auxiliary circuits include a phase generator, switch drivers, Conversion Ratio (CR) controller, cell controller, voltage reference generator, and start-up circuits.

Chapter 6 covers the implementation of the 16 mW PMU implementation in 130 nm CMOS bulk technology, to evaluate its performance and validate the theoretical equations through schematic simulations. The PCB test board and the test setup used for the circuit evaluation are also described. The measurement results of the prototype are presented, compared with the simulation ones, and conclusions are drawn.

Finally, in Chap. 7 conclusions are drawn and future work is discussed.

References

1. Smith IG (2012) The internet of things 2012 new horizons. European Research Cluster on the Internet of Things (IERC), UK
2. Kortuem G, Kawsar F, Sundramoorthy V, Fitton D (2010) Smart objects as building blocks for the internet of things. IEEE Internet Comput. https://doi.org/10.1109/MIC.2009.143
3. Myers J, Savanth A, Howard D, Gaddh R, Prabhat P, Flynn D (2015) An 80 nW retention 11.7 pJ/cycle active subthreshold ARM Cortex-M0+ subsystem in 65 nm CMOS for WSN applications. In: 2015 IEEE international solid-state circuits conference - (ISSCC) Digest of technical papers. https://doi.org/10.1109/ISSCC.2015.7062967
4. Klinefelter A, Roberts NE, Shakhsheer Y, Gonzalez P, Shrivastava A, Roy A, Craig K, Faisal M, Boley J, Seunghyun O, Yanqing Z, Akella D, Wentzloff DD, Calhoun BH (2015) A 6.45 μW self-powered IoT SoC with integrated energy-harvesting power management and ULP asymmetric radios. In: 2015 IEEE international solid-state circuits conference (ISSCC). https://doi.org/10.1109/ISSCC.2015.7063087
5. Omairi A, Ismail ZH, Danapalasingam KA, Ibrahim M (2017) Power harvesting in wireless sensor networks and its adaptation with maximum power point tracking: current technology and future directions. IEEE Internet Things J. https://doi.org/10.1109/JIOT.2017.2768410
6. Liu X, Sánchez-Sinencio E (2015) An 86% efficiency 12 μW self-sustaining PV energy harvesting system with hysteresis regulation and time-domain MPPT for IoT smart nodes. IEEE J Solid-State Circuits. https://doi.org/10.1109/JSSC.2015.2418712
7. Liu X, Huang L, Ravichandran K, Sánchez-Sinencio E (2016) A highly efficient reconfigurable charge pump energy harvester with wide harvesting range and two-dimensional MPPT for internet of things. IEEE J Solid-State Circuits. https://doi.org/10.1109/JSSC.2016.2525822
8. Rawy K, Yoo T, Kim TT (2018) An 88% efficiency 0.1–300-μ W energy harvesting system with 3-D MPPT using switch width modulation for iot smart nodes. IEEE J Solid-State Circuits. https://doi.org/10.1109/JSSC.2018.2833278
9. Jiang J, Lu Y, Huang C, Ki W, Mok PKT (2015) A 2-/3-phase fully integrated switched-capacitor DC-DC converter in bulk CMOS for energy-efficient digital circuits with 14% efficiency improvement. In: 2015 IEEE international solid-state circuits conference (ISSCC). https://doi.org/10.1109/ISSCC.2015.7063078
10. Carreon-Bautista S, Huang L, Sánchez-Sinencio E (2016) An autonomous energy harvesting power management unit with digital regulation for IoT applications. IEEE J Solid-State Circuits. https://doi.org/10.1109/JSSC.2016.2545709
11. Liu X, Ravichandran K, Sánchez-Sinencio E (2018) A switched capacitor energy harvester based on a single-cycle criterion for MPPT to eliminate storage capacitor. IEEE Trans Circuits Syst I Regul Pap. https://doi.org/10.1109/TCSI.2017.2726345
12. Weddell AS, Merrett GV, Kazmierski TJ, Al-Hashimi BM (2011) Accurate supercapacitor modelling for energy harvesting wireless sensor nodes. IEEE Trans Circuits Syst II, Exp Briefs. https://doi.org/10.1109/TCSII.2011.2172712

13. Hua X, Harjani R (2015) 3.5–0.5 V input, 1.0 V output multi-mode power transformer for a supercapacitor power source with a peak efficiency of 70.4%. In: 2015 IEEE Custom Integrated Circuits Conference (CICC). https://doi.org/10.1109/CICC.2015.7338390

14. Sudevalayam S, Kulkarni P (2011) Energy harvesting sensor nodes: survey and implications. IEEE Commun Surv Tutor. https://doi.org/10.1109/SURV.2011.060710.00094

15. Yang H, Zhang Y (2013) Analysis of supercapacitor energy loss for power management in environmentally powered wireless sensor nodes. IEEE Trans Power Electron. https://doi.org/10.1109/TPEL.2013.2238683

16. Carvalho C, Lavareda G, Lameiro J, Paulino N (2011) A step-up μ-power converter for solar energy harvesting applications, using Hill Climbing maximum power point tracking. In: IEEE International Symposium of Circuits and Systems (ISCAS). https://doi.org/10.1109/ISCAS.2011.5937965

17. Ozaki T, Hirose T, Asano H, Kuroki N, Numa M (2016) Fully-integrated high-conversion-ratio dual-output voltage boost converter with MPPT for low-voltage energy harvesting. IEEE J Solid-State Circuits. https://doi.org/10.1109/JSSC.2016.2582857

18. Sarafianos A, Steyaert M (2015) Fully integrated wide input voltage range capacitive DC-DC converters: the folding dickson converter. IEEE J Solid-State Circuits. https://doi.org/10.1109/JSSC.2015.2410800

19. Lu Y, Jiang J, Ki W (2017) A multiphase switched-capacitor DC-DC converter ring with fast transient response and small ripple. IEEE J Solid-State Circuits. https://doi.org/10.1109/JSSC.2016.2617315

20. Jiang Y, Law M, Chen Z, Mak P, Martins RP (2019) Algebraic series-parallel-based switched-capacitor DC-DC boost converter with wide input voltage range and enhanced power density. IEEE J Solid-State Circuits. https://doi.org/10.1109/JSSC.2019.2935556

21. Rincon-Mora GA, Allen PE (1998) A low-voltage, low quiescent current, low drop-out regulator. IEEE J Solid-State Circuits. https://doi.org/10.1109/4.654935

22. Milliken JR, Silva-Martinez J, Sanchez-Sinencio E (2007) Full on-chip cmos low-dropout voltage regulator. IEEE Trans Circuits Syst I Regul Pap. https://doi.org/10.1109/TCSI.2007.902615

23. Cheung FL, Mok PKT (2004) A monolithic current-mode CMOS DC-DC converter with on-chip current-sensing technique. IEEE J Solid-State Circuits. https://doi.org/10.1109/JSSC.2003.820870

24. Wibben J, Harjani R (2008) A high-efficiency DC-DC converter using 2 nH integrated inductors. IEEE J Solid-State Circuits. https://doi.org/10.1109/JSSC.2008.917321

25. Forouzesh M, Shen Y, Yari K, Siwakoti YP, Blaabjerg F (2018) High-efficiency high step-up DC-DC converter with dual coupled inductors for grid-connected photovoltaic systems. IEEE Trans Power Electron. https://doi.org/10.1109/TPEL.2017.2746750

26. Salvador MA, Lazzarin TB, Coelho RF (2018) High step-up DC-DC converter with active switched-inductor and passive switched-capacitor networks. IEEE Trans Industr Electron. https://doi.org/10.1109/TIE.2017.2782239

27. Amin SS, Mercier PP (2019) A fully integrated Li-Ion-compatible hybrid four-level DC-DC converter in 28-nm FDSOI. IEEE J Solid-State Circuits. https://doi.org/10.1109/JSSC.2018.2880183

28. Andersen TM, Krismer F, Kolar JW, Toifl T, Menolfi C, Kull L, Morf T, Kossel M, Brändli M, Buchmann P, Francese PA (2014). IEEE international solid-state circuits conference digest of technical papers (ISSCC). https://doi.org/10.1109/ISSCC.2014.6757351

29. Hua Z, Lee H (2015) A reconfigurable dual-output switched-capacitor DC-DC regulator with sub-harmonic adaptive-on-time control for low-power applications. IEEE J Solid-State Circuits. https://doi.org/10.1109/JSSC.2014.2379616

30. Jianxi L, Hao C, Tianyuan H, Junchao M, Zhangming Z, Yintang Y (2018) A dual mode step-down switched-capacitor DC-DC converter with adaptive switch width modulation. Microelectron J. https://doi.org/10.1016/j.mejo.2018.06.003

31. Jiang Y, Law M, Mak P, Martins RP (2018) Algorithmic voltage-feed-in topology for fully integrated fine-grained rational Buck-Boost switched-capacitor DC-DC converters. IEEE J Solid-State Circuits. https://doi.org/10.1109/JSSC.2018.2866929

32. Jiang J, Liu X, Huang C, Ki W, Mok PKT, Lu Y (2020) Subtraction-mode switched-capacitor converters with parasitic loss reduction. IEEE Trans Power Electron. https://doi.org/10.1109/TPEL.2019.2933623

33. Wens M, Steyaert M (2013) Design and implementation of fully-integrated inductive DC-DC converters in standard CMOS. Analog circuits and signal processing. Springer, Netherlands

34. Jiang J, Liu X, Mok Ki WH, PKT, Lu Y (2021) Circuit techniques for high efficiency fully-integrated switched-capacitor converters. IEEE Trans Circuits Syst II: Express Briefs. https://doi.org/10.1109/TCSII.2020.3046514

35. Le H, Sanders SR, Alon E (2011) Design techniques for fully integrated switched-capacitor DC-DC converters. IEEE J Solid-State Circuits. https://doi.org/10.1109/JSSC.2011.2159054

Switched-Capacitor DC-DC Fundamentals

2

This chapter describes the working principle of Switched Capacitors (SCs) DC-DC converters. It includes a detailed analysis of the main components of these type of converters, capacitors and switches, and its implementation in Complementary Metal-Oxide-Semiconductor (CMOS) technology. It also describes a step-by-step design methodology, applied as an example to a Series-Parallel (SP) step-down converter with a Conversion Ratio (CR) of 1/2. Lastly, an overview of the loop regulation techniques is briefly described.

2.1 Capacitor Implementation in CMOS

In this section, three of the most common structures used to implement capacitors in typical CMOS technologies are characterized. These are the CMOS, Metal-Insulator-Metal (MIM), and Metal-Oxide-Metal (MOM) capacitors.

2.1.1 Metal-Oxide-Semiconductor Capacitors

Figure 2.1a shows the PMOS transistor configuration to implement a capacitor, known as MOS capacitor [1, 2]. In this configuration, the transistor has two terminals—the gate and the source, drain, and bulk, all shorted together. NMOS transistors can also be used to implement a MOS capacitor; however, the MOS capacitor requires the channel to be isolated from the substrate, for device's well to remain reversed biased, to decrease the capacitance parasitics [3, 4]. Which in the NMOS requires using a triple-well device, which has a larger parasitic capacitance when compared to the PMOS device, which is naturally isolated. The reverse-biasing pn-junction to ground, on the PMOS capacitor, results in a junction capacitance between the n-well and ground (p-sub), meaning that the PMOS capacitor has a

© The Author(s), under exclusive license to Springer Nature Switzerland AG 2022
R. Madeira et al., *Fully Integrated Switched-Capacitor PMU for IoT Nodes*, Synthesis
Lectures on Engineering, Science, and Technology,
https://doi.org/10.1007/978-3-031-14701-2_2

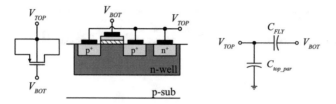

(a) Schematic and PMOS cross-sectional view.   (b) Equivalent model.

Fig. 2.1 PMOS capacitor simplified and equivalent schematic

Fig. 2.2 Simulation results:
PMOS capacitor capacitance
density as a function of V_{SG}

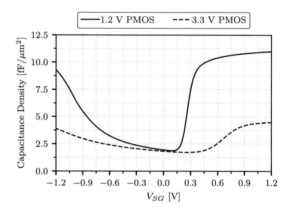

top parasitic capacitance, whose value can reach up to 7% [5–7] of the nominal capacitance value when in the strong-inversion region. The C_{GB} capacitance can be neglected in the inversion region; thus, there is no significant bottom parasitic capacitance. The simplified model of the PMOS capacitor is shown in Fig. 2.1b.

Figure 2.2 shows the plot of the capacitance density as a function of V_{SG} for 1.2 and 3.3 V PMOS capacitor, in 130 nm CMOS technology. The maximum capacitance value is achieved in either the accumulation region ($V_{SG} < 0$) or the inversion region ($V_{SG} > |V_{th}|$). However, in the accumulation region, the dominant capacitance is the gate-to-bulk capacitance, C_{GB}, which has a significant series parasitic resistance, coming from the physical distance between the substrate connection and the substrate region under the oxide. Hence, it is preferable to operate the MOS capacitor in the inversion region [8]. In the inversion region, the capacitance density ($C_{den} = C_{nom}/A_c$) is approximately 10–11 fF/μm^2, for a 1.2 V PMOS transistor, in 130 nm bulk CMOS technology, where C_{nom} is the nominal capacitance value and A_c is the total capacitance area. Due to the thicker oxide, the 3.3 V PMOS capacitor has a lower capacitance density, approximately less than half of the 1.2 V PMOS value. Hence, it is important to keep the voltage across the capacitor under the 1.2 V PMOS voltage rating limits so that the capacitance density can be maximized.

The series resistance of the MOS capacitance arises from the resistivity of the gate material and the channel resistance. The dominant resistance is the channel resistance,

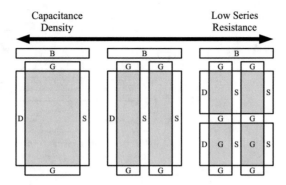

Fig. 2.3 MOS capacitor layout and its influence on capacitance density and series resistance, adapted from [1, 3]

assuming that the layout is designed in a way that minimizes the gate resistance, e.g. by minimizing the poly connection and by using a large number of vias. Has shown in [1], the channel resistance is proportional to the transistor's length. Hence, wide and multiple finger transistors are preferred to minimize the series resistance. However, the overhead area of the drain/source/bulk contacts decreases the capacitance density, as shown in Fig. 2.3.

2.1.2 Metal-Insulator-Metal Capacitors

The MIM capacitor is formed by two metal planes in the upper layers separated by a thin insulator film (Fig. 2.4a). Since these capacitors are implemented in the upper metal layers, their distance to the substrate is relatively high, reducing the capacitive coupling between the capacitor plates and the substrate, thus having low parasitic values, typically around 1–2% [3, 6, 7, 9]. These metal plates are asymmetric—the bottom plate is larger than the top plate, and since it is closer to the substrate, it results in a bottom parasitic capacitance larger than the top parasitic capacitance ($C_{Bot} > C_{Top}$).

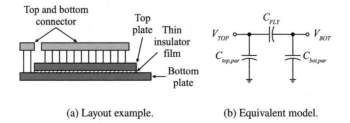

(a) Layout example.　　　　　(b) Equivalent model.

Fig. 2.4 Simplified layout cross-section example and equivalent model of a MIM capacitor

Fig. 2.5 Simulation results: capacitance density as a function of V_C for a MIM and MOM capacitor

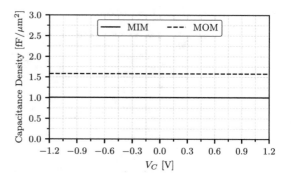

Figure 2.5 shows the simulation results of the capacitance density as a function of the capacitor's voltage for the MIM capacitor, in 130 nm CMOS technology. The capacitance density value is constant throughout the voltage range of -1.2–1.2 V, and has a capacitance density of approximately 1 fF/μm^2.

2.1.3 Metal-Oxide-Metal Capacitors

The MOM capacitors, similarly to the MIM are also implemented in the metal layers, however, instead of a large solid plate, like the MIM capacitors, MOM capacitors plates are broken into fingers, resulting in metal fingers running in parallel, creating coupling capacitors that increase the capacitance. Figure 2.6 shows a simplified diagram of a symmetrical and asymmetrical MOM capacitor. The asymmetrical has a dominant bottom parasitic capacitance, whilst the symmetrical has both the top and bottom parasitic capacitance with approximately the same value ($\approx$1.5% [7]). Notice that both the top and bottom plates are implemented in the same metal layer. Furthermore, the metal layers can be stacked from the bottom to the top, in 130 nm from metal 1 to metal 6, increasing even further the capacitance density. However, since the metal layers are closer to the substrate, the parasitic capacitances coupled through the substrate are higher than the MIM capacitors [9].

Figure 2.5 shows the simulation results of the capacitance density as a function of the capacitor's voltage for the MOM capacitor, in 130 nm CMOS technology. The capacitance value does not depend on the capacitor's voltage value as shown in the plot and has a capacitance density of approximately 1.6 fF/μm^2.

2.1.4 Exotic Capacitors

The capacitors implementations described above are available in bulk CMOS technology. However, there are other capacitor implementations, which use extra masks in the fabrication process, that can provide higher capacitance density and lower parasitic capacitance values.

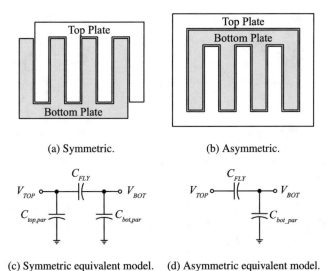

(a) Symmetric.   (b) Asymmetric.

(c) Symmetric equivalent model.   (d) Asymmetric equivalent model.

Fig. 2.6 Simplified layout example and equivalent model of a MOM capacitor

In the literature, ferroelectric capacitors (Fe-Caps) [10] and deep trench capacitors, available in Silicon-On-Insulator (SOI) CMOS technology [11–15], were used to achieve higher power density and reaching higher peak efficiency values. Nonetheless, this comes at the cost of increasing the fabrication price. The implementation of these capacitors will not be further detailed, since one of the main advantages of fully integrating an SC converter is the price reduction, and thus this work focus on native capacitor implementations in standard CMOS technology.

2.2   Switch Implementation

As shown in Fig. 2.7, an analog switch can be implemented by a single NMOS or PMOS transistor, or by combining both in parallel (Transmission Gate). Figure 2.8 shows the simulated ON-Resistance (R_{ON}) of the three mentioned switch structures as a function of V_{IN}, for $V_G = 0.9$ V in the NMOS case and for $V_G = 0$ V in the PMOS case. The value of R_{ON} is only low for low V_{IN} values, in the case of the NMOS transistor, and for high V_{IN} values, in the case of the PMOS transistor. If low R_{ON} is required throughout the whole V_{IN} range, then the switch must be implemented by a transmission gate. Hence, the type of device chosen depends on the voltage at the switch terminals when it is closed.

For small drain-to-source voltages ($V_{DS} \ll V_{GS} - V_{TH}$), the switch R_{ON} and C_{GG} are given by

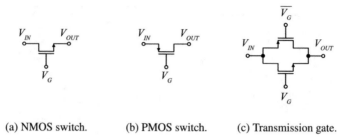

(a) NMOS switch.   (b) PMOS switch.   (c) Transmission gate.

Fig. 2.7 Transistor implementations of an analogue switch, PMOS and NMOS with undefined bulk have their bulk connected do V_{Source} and V_{SS}, respectively

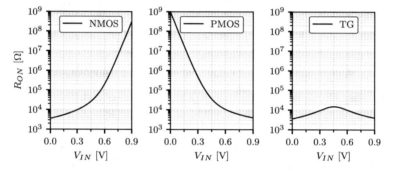

Fig. 2.8 Simulation results: transistor's R_{ON} as a function of V_{IN} for an NMOS, PMOS, and transmission gate

$$R_{ON} \cong \frac{L}{C_{ox}\mu_n W(V_{GS} - V_{TH})} \approx \frac{k_R}{W} \Rightarrow k_R = R_{ON}\,W\ (\Omega \cdot m) \tag{2.1}$$

$$C_{GG} = C_{GD} + C_{GS} \cong WLC_{ox} + WC_{ov} \approx k_C\,W \Rightarrow k_C = \frac{C_{GG}}{W}\ (F/m) \tag{2.2}$$

These equations show that R_{ON} is inversely proportional to the transistor's width (W), and that C_{GG} is directly proportional to W. Furthermore, if V_{GS}, L, C_{ox} and μ_n are kept constant, and $V_{DS} \ll V_{GS} - V_{TH}$, then the previous equations can be approximated by a constant coefficient, k_R and k_C, that relates both R_{ON} and C_{GG} with the transistor's width (W). These coefficients can be derived from electric simulations, by using the test bench shown in Fig. 2.9. This work was implemented using the 130 nm CMOS bulk technology and four transistors were tested—the NMOS 1.2 and 3.3 V, and the PMOS 1.2 and 3.3 V. The 1.2 V transistors have a thinner oxide than the 3.3 V transistors. All the four transistors were tested using the minimum length allowed by the technology, and the voltages values used in the simulations for the NMOS were $V_{GS} = 0.9$ V and $V_{DS} = 5$ mV. For the PMOS, the values used were $V_{SG} = 0.9$ V, $V_S = 0.9$ V and $V_{DS} = 5$ mV.

Table 2.1 shows the k_R and k_C values for a 130 nm CMOS technology for the conditions mentioned in the previous paragraph. It can be seen that the values of k_R are much higher

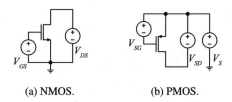

(a) NMOS.   (b) PMOS.

Fig. 2.9 Schematic test bench for measuring the k_R and k_C constants

Table 2.1 130 nm technology NMOS and PMOS k_R and k_C values, for $V_{GS} = 0.9$ V

Device	NMOS		PMOS	
Voltage rating (V)	**1.2**	**3.3**	**1.2**	**3.3**
k_R (k$\Omega \cdot \mu$m)	0.58	5.34	2.71	20.57
k_C (fF/μm)	1.34	1.77	1.41	1.95

Fig. 2.10 Simulation results: switch C_{GG} as a function of R_{ON} for 130 nm CMOS technology, for $V_{GS} = 0.9$ V

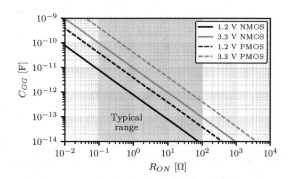

for the 3.3 V transistors than for the 1.2 V, around 9.2 times for the NMOS and 7.6 times for the PMOS transistor. Hence, it is preferable to select 1.2 V transistors. Moreover, the NMOS transistor has a k_R coefficient 4.7 times smaller than the PMOS k_R value. Therefore, when possible, the 1.2 V NMOS transistor should be used. The k_C values are closer to each other than the k_R values, which at first glance could give the impression that the C_{GG} value does not vary much. However, since C_{GG} depends on the multiplication of both k_R and k_C, its value increases significantly in the 3.3 V transistors. Figure 2.10 shows C_{GG} as a function of R_{ON}, for both NMOS and PMOS transistors.

2.3 Theoretical SC DC-DC Analysis

Figure 2.11 shows the schematic of a step-down SP SC converter with a CR of 1/2. It is composed of a flying capacitor C_{FLY} and four switches, where $\phi_{1,2}$ are two clock signals complementary to each other with 50% duty cycle. Hence, during phase ϕ_1, C_{FLY} connects

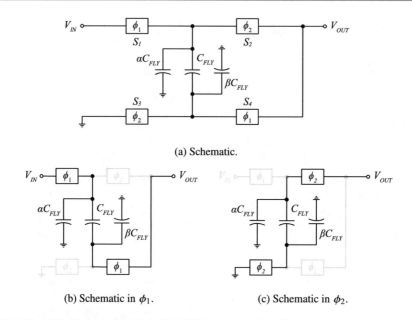

(a) Schematic.

(b) Schematic in ϕ_1.   (c) Schematic in ϕ_2.

Fig. 2.11 Simplified schematic of the SP 1/2 SC converter [16, 17]

between the input voltage (V_{IN}) and the output voltage (V_{OUT}) and, during phase ϕ_2, C_{FLY} connects between V_{OUT} and ground. This is shown in Fig. 2.11a. The C_{FLY} parasitic capacitances are represented in the schematic by α and β. These refer to the top and bottom parasitic capacitance, respectively, as percentage of the nominal C_{FLY} value.

Assuming that V_{OUT} is kept at a constant voltage, the charge conservation equations are given by

$\phi_1 \rightarrow \phi_2$:

$$(V_{IN} - V_{OUT})\, C_{FLY} + V_{IN}\, (\alpha\, C_{FLY}) = V_{OUT}\, (C_{FLY} + \alpha\, C_{FLY}) + \Delta q_o^{\phi_2} \tag{2.3}$$

$\phi_2 \rightarrow \phi_1$:

$$- V_{OUT}\, C_{FLY} = (V_{OUT} - V_{IN})\, C_{FLY} + V_{OUT}\, (\beta\, C_{FLY}) + \Delta q_o^{\phi_1} \tag{2.4}$$

$$V_{OUT}\, (C_{FLY} + \alpha\, C_{FLY}) = (V_{IN} - V_{OUT})\, C_{FLY} + V_{IN}\, (\alpha\, C_{FLY}) - \Delta q_i^{\phi_1} \tag{2.5}$$

where $\Delta q_o^{\phi_{1,2}}$ are the amount of charge absorbed by V_{OUT}, in the respective phase, and $\Delta q_i^{\phi_1}$ is the amount of charge drawn by the circuit from V_{IN}, in this case only during ϕ_1. These equations can be solved in respect to $\Delta q_i^{\phi_1}$, $\Delta q_o^{\phi_1}$, and $\Delta q_o^{\phi_2}$, which results in

$$\Delta q_i^{\phi_1} = C_{FLY} \left(V_{IN} \left(1 + \alpha \right) - V_{OUT} \left(2 + \alpha \right) \right) \tag{2.6}$$

$$\Delta q_o^{\phi_1} = C_{FLY} \left(V_{IN} - V_{OUT} \left(2 + \beta \right) \right) \tag{2.7}$$

$$\Delta q_o^{\phi_2} = C_{FLY} \left(V_{IN} \left(1 + \alpha \right) - V_{OUT} \left(2 + \alpha \right) \right) \tag{2.8}$$

The input and output current and power can then be determined by

$$I_{IN} = \Delta q_i^{\phi_1} F_{CLK} = C_{FLY} F_{CLK} \left(V_{IN} \left(1 + \alpha \right) - V_{OUT} \left(2 + \alpha \right) \right) \tag{2.9}$$

$$I_{OUT} = \left(\Delta q_o^{\phi_1} + \Delta q_o^{\phi_2} \right) F_{CLK} = C_{FLY} F_{CLK} \left(V_{IN} \left(2 + \alpha \right) - V_{OUT} \left(4 + \alpha + \beta \right) \right) \tag{2.10}$$

$$P_{IN} = V_{IN} I_{IN} = C_{FLY} F_{CLK} V_{IN} \left(V_{IN} \left(1 + \alpha \right) - V_{OUT} \left(2 + \alpha \right) \right) \tag{2.11}$$

$$P_{OUT} = V_{OUT} I_{OUT} = C_{FLY} F_{CLK} V_{OUT} \left(V_{IN} \left(2 + \alpha \right) - V_{OUT} \left(4 + \alpha + \beta \right) \right) \tag{2.12}$$

The converter efficiency (η) can be obtained by

$$\eta = \frac{P_{OUT}}{P_{IN}} = \frac{V_{OUT} \left(V_{IN} \left(2 + \alpha \right) - V_{OUT} \left(4 + \alpha + \beta \right) \right)}{V_{IN} \left(V_{IN} \left(1 + \alpha \right) - V_{OUT} \left(2 + \alpha \right) \right)} \tag{2.13}$$

The converter can be modelled as an ideal transformer [18, 19, 21] as shown in Fig. 2.12, where the load line plot (I_{OUT} vs V_{OUT}) and the ideal efficiency plot (η vs V_{OUT}) are also depicted. It can be seen that when V_{OUT} equals V_{MAX} the output current is zero and increases linearly as V_{OUT} increases. Hence, R_{OUT} can be determined by (2.14). Furthermore, the CR can also be determined by dividing V_{max} by V_{IN}, or I_{IN} by I_{OUT}, as shown in (2.15).

$$R_{OUT} \Big|_{\alpha, \beta = 0} = \frac{V_{MAX} - V_{OUT}}{I_{OUT}} = \frac{1}{4 \, C_{FLY} \, F_{CLK}} \tag{2.14}$$

$$\text{CR} \Big|_{\alpha, \beta = 0} = \frac{V_{MAX}}{V_{IN}} = \frac{I_{IN}}{I_{OUT}} = \frac{1}{2} \tag{2.15}$$

Fig. 2.12 Ideal transformer model, load line and ideal efficiency of an SC converter [18]

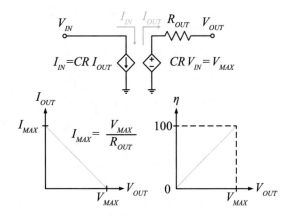

Finally, Considering a load resistor (R_L) connected to the output, the output voltage (V_{OUT}) can be determining using Ohm's law (2.16), and the converter's clock frequency F_{CLK} can be obtained by solving (2.16) for F_{CLK} (2.17). Knowing that $P_{OUT} = V_{OUT}^2/R_L$, F_{CLK} can be also given as a function of P_{OUT} (2.18).

$$V_{OUT} = I_{OUT} R_L \Rightarrow V_{OUT} = \frac{C_{FLY} \, F_{CLK} \, R_L \, V_{IN} \, (2 + \alpha)}{1 + C_{FLY} \, F_{CLK} \, R_L \, (4 + \alpha + \beta)} \tag{2.16}$$

$$F_{CLK} = \frac{V_{OUT}}{C_{FLY} \, R_L \, (V_{IN} \, (2 + \alpha) - V_{OUT} \, (4 + \alpha + \beta))} \tag{2.17}$$

$$F_{CLK} = \frac{P_{OUT}}{C_{FLY} \, V_{OUT} \, (V_{IN} \, (2 + \alpha) - V_{OUT} \, (4 + \alpha + \beta))} \tag{2.18}$$

Figure 2.13 shows the converter's efficiency and F_{CLK} as a function of V_{IN} for different values of α and β, with $C_{FLY} = 100$ pF and $P_{OUT} = 1$ mW. The plots clearly show that both parasitic capacitances have a negative impact on both efficiency and frequency. However, the top parasitic capacitance has a smaller impact, when compared with the bottom parasitic

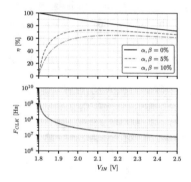

(a) Top and bottom parasitic capacitances impact.

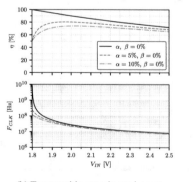

(b) Top parasitic capacitance impact.

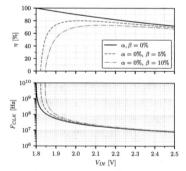

(c) Bottom parasitic capacitance impact.

Fig. 2.13 Converter's efficiency as a function of V_{IN}, for $V_{OUT} = 1$ V, $C_{FLY} = 100$ pF, and $P_{OUT} = 1$ mW

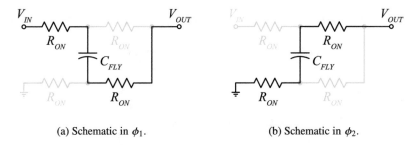

(a) Schematic in ϕ_1.   (b) Schematic in ϕ_2.

Fig. 2.14 Simplified schematic of the SP 1/2 SC converter with the switches replaced by R_{ON}

capacitance, and it pushes the peak efficiency for lower V_{IN} levels whilst decreasing F_{CLK} for the same input/output ratio. This is because the charge absorbed during ϕ_1 is then supplied to V_{OUT} during ϕ_2. This acts like a parallel 1/1 converter, and thus, allows the converter to work at a lower frequency for the same input/output voltage ratio. Hence, in this topology, when implementing C_{FLY} the plate with the highest parasitic capacitance should be used as the capacitor top plate.

Until now, the efficiency of the converter has been calculated assuming that the clock phases are long enough to allow C_{FLY} to completely charge (or discharge). However, the finite R_{ON} value of the switches may cause partial charging, depending on the frequency of operation value [12, 18, 19, 21]. Figure 2.14 shows the converter's schematic with its ON switches replaced by their respective R_{ON}.

Considering the switches' finite resistance, R_{ON}, the C_{FLY} voltage in each phase can be re-written as

$$V_{C_{FLY}}^{\phi_1} = (V_{IN} - V_{OUT}) + \left(V_{final}^{\phi_2} - (V_{IN} - V_{OUT})\right) e^{\left(\frac{-1}{2\,R_{ON}\,C_{FLY}\,F_{CLK}}\right)} \qquad (2.19)$$

$$V_{C_{FLY}}^{\phi_2} = V_{OUT} + \left(V_{final}^{\phi_1} - V_{OUT}\right) e^{\left(\frac{-1}{2\,R_{ON}\,C_{FLY}\,F_{CLK}}\right)} \qquad (2.20)$$

where $V_{final}^{\phi_1}$ and $V_{final}^{\phi_2}$ are the C_{FLY} voltages at the end of ϕ_1 and ϕ_2, respectively. In ϕ_1, $V_{final}^{\phi_1}$ corresponds to the maximum C_{FLY} voltage, and in ϕ_2, $V_{final}^{\phi_2}$ corresponds to the minimum C_{FLY} voltage. Hence, Eqs. (2.20) and (2.19) can be re-written as:

$$V_{C_{FLY\,MAX}} = (V_{IN} - V_{OUT}) + \left(V_{C_{FLY\,MIN}} - (V_{IN} - V_{OUT})\right) e^{\left(\frac{-1}{2\,R_{ON}\,C_{FLY}\,F_{CLK}}\right)} \qquad (2.21)$$

$$V_{C_{FLY\,MIN}} = V_{OUT} + \left(V_{C_{FLY\,MAX}} - V_{OUT}\right) e^{\left(\frac{-1}{2\,R_{ON}\,C_{FLY}\,F_{CLK}}\right)} \qquad (2.22)$$

Solving the above equations for $V_{C_{FLY_{MAX}}}$ and $V_{C_{FLY_{MIN}}}$, $V_{C_{FLY}}$ voltage variation is given by

$$\Delta V_{C_{FLY}} = V_{C_{FLY_{MAX}}} - V_{C_{FLY_{MIN}}} = \gamma \ (V_{IN} - 2\ V_{OUT}) \tag{2.23}$$

where

$$\gamma = \left(\frac{1 - e^{\left(\frac{-1}{2\ R_{ON}\ C_{FLY}\ F_{CLK}} \right)}}{1 + e^{\left(\frac{-1}{2\ R_{ON}\ C_{FLY}\ F_{CLK}} \right)}} \right) \tag{2.24}$$

As shown in Fig. 2.14, C_{FLY} connects to the output node in both phases. Hence, C_{FLY} delivers an amount of charge of $C_{FLY}\ \Delta V_{C_{FLY}}$ in each phase, giving a total charge given to the output of (2.25) [12].

$$Q_{C_{FLY}} = 2\ C_{FLY}\ \Delta V_{C_{FLY}} = 2\ \gamma\ C_{FLY}\ (V_{IN} - 2\ V_{OUT}) \tag{2.25}$$

Thus, considering $\alpha = \beta = 0\%$, the I_{OUT} (2.10) and R_{OUT} (2.14) equations can be re-written as

$$I_{OUT} = Q_{C_{FLY}}\ F_{CLK} = 2\ \gamma\ C_{FLY}\ F_{CLK}\ (V_{IN} - 2\ V_{OUT}) \tag{2.26}$$

$$R_{OUT} = \frac{1}{\gamma}\ \frac{1}{4\ C_{FLY}\ F_{CLK}} \tag{2.27}$$

These equations show that Eqs. (2.26) and (2.10) are identical except for a scaling factor, γ, which accounts for incomplete charging [12, 18]. Let N be the number of time constants, defined as

$$N = \frac{1}{2\ R_{ON}\ C_{FLY}\ F_{CLK}} \tag{2.28}$$

then γ can be given as a function of N, where a large value of N, results in a fast settling time and a low value results in a slow settling time. Being N inversely proportional to $R_{ON}\ C_{FLY}$, for a given C_{FLY}, increasing the settling time comes at the cost of having low R_{ON} values, resulting in large transistor area, and thus, the higher power dissipation of the switches. Knowing that the error between steady-state value and the value at the end of the phase $(1/(2\ F_{CLK}))$ increases as N decreases, there is a trade-off between the error and N. Table 2.2 shows that the value of γ drops significantly for values lower than $N = 4$. Figure 2.15 shows R_{OUT} for $N = 3, 4, 5$ as a function of F_{CLK} with (2.27) and without (2.14) the switches finite R_{ON} impact, for $C_{FLY} = 100$ pF, and F_{CLK} calculated using (2.18), neglecting the parasitic capacitance effects ($\alpha = \beta = 0\%$), for $V_{IN} = 1.9$ V, $V_{OUT} = 0.9$ V and $P_{OUT} = 1$ mW. The figure shows that, at the operating frequency F_{CLK}, using $N = 3$ significantly deviates from the ideal R_{OUT} value when compared with $N = 4$ or $N = 5$. Since $N = 4$ and $N = 5$ are very similar, the first, $N = 4$ offers a good compromise point for sizing the converter switches.

Table 2.2 γ values for different values of the number of time constants (N)

N	2	3	4	5	6
γ (%)	76.16	90.51	96.40	98.66	99.51

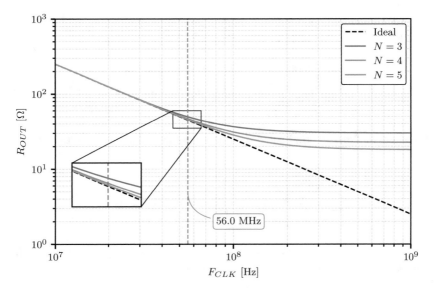

Fig. 2.15 Converter's R_{OUT} as a function of F_{CLK}, with the switches finite R_{ON} impact

Fixing $N = 4$ allows to size the switches R_{ON} and thus determining the transistor's W. Hence, the power dissipation of charging the gate switches' parasitic capacitance (C_{GG}) can be determined, which will also impact the converter's efficiency. Other switches parasitic capacitances, like the source (C_{SS}) and drain (C_{DD}) nodes, are in parallel with either C_{FLY} of decoupling capacitors, which are typically much larger than C_{DD} or C_{SS}, and thus can be neglected.

Figure 2.14 shows the converter's schematic with its ON switches replaced by the respective R_{ON}. Both ϕ_1 and ϕ_2 (Fig. 2.14a and b, respectively) have the same time constant given by

$$\tau = R_{ON_{tot}} C_{FLY} = \frac{1}{2 N F_{CLK}} \tag{2.29}$$

where $R_{ON_{tot}} = R_{ON_{S1}} + R_{ON_{S2}}$ is the total resistance value of switches in series per phase. Assuming that $R_{ON_{S1}} = R_{ON_{S2}} = R_{ON}$, which means that $R_{ON} = R_{ON_{tot}}/2$, then, the R_{ON} value for the $N = 4$ is given by

$$R_{ON_{tot}} = \frac{1}{4 \times 2 \, C_{FLY} \, F_{CLK}} \Rightarrow R_{ON} = \frac{1}{16 \, C_{FLY} \, F_{CLK}} \tag{2.30}$$

The capacitance C_{GG} can be determined by combining the previous equation with (2.1) and (2.2), as shown in (2.31). The switches' power dissipation can be calculated by multiplying the sum of all the switches' C_{GG}, the F_{CLK}, and the switches gate-to-source voltage swing V_{SW} squared, as shown in Eq. (2.34), where $K_{SW} = k_{C_1} k_{R_1} + k_{C_2} k_{R_2} + \cdots + k_{C_N} k_{R_N}$ [34].

$$C_{GG} = \frac{k_R k_C}{R_{ON}} = 16 k_C k_R C_{FLY} F_{CLK} \tag{2.31}$$

$$P_{SW} = (C_{GG_{S1}} + C_{GG_{S2}} + C_{GG_{S3}} + C_{GG_{S4}}) F_{CLK} V_{SW}^2 = \tag{2.32}$$

$$= 16 (k_{C_1} k_{R_1} + k_{C_2} k_{R_2} + \cdots + k_{C_N} k_{R_N}) C_{FLY} F_{CLK}^2 V_{SW}^2 = \tag{2.33}$$

$$= 16 K_{SW} C_{FLY} F_{CLK}^2 V_{SW}^2 \tag{2.34}$$

The effect of P_{SW} can now be added to the converter's efficiency, resulting in

$$\eta = \frac{P_{OUT}}{P_{IN} + P_{SW}} = \tag{2.35}$$

$$= \frac{V_{OUT} (V_{IN} (\alpha + 2) - V_{OUT} (\alpha + \beta + 4))}{16 F_{CLK} V_{OUT}^2 K_{SW} + V_{IN}^2 (\alpha + 1) - V_{IN} V_{OUT} (\alpha + 2)} \tag{2.36}$$

where for simplicity it is assumed that $V_{SW} = V_{OUT}$. Due to the F_{CLK}^2 term in (2.34), F_{CLK} does not cancel out. Thus, replacing F_{CLK} by (2.18) yields

$$\eta = \frac{V_{OUT} (V_{IN} (\alpha + 2) - V_{OUT} (\alpha + \beta + 4))}{V_{IN}^2 (\alpha + 1) - V_{IN} V_{OUT} (\alpha + 2) + \dfrac{16 K_{SW} P_{OUT} V_{OUT}}{C_{FLY} (V_{IN} (2 + \alpha) - V_{OUT} (4 + \alpha + \beta))}} \tag{2.37}$$

The capacitance C_{FLY} can be replaced by the capacitance's area times the capacitance density (C_{den}) of the device chosen to implement it, e.g. ≈ 10 fF/μm^2 for the MOS capacitor. Then, C_{FLY} can be re-written by $C_{FLY} = A_c \times C_{den}$. Replacing this in (2.37), gives the efficiency as a function of P_{OUT} per capacitance area, i.e. power density, as shown below.

$$\eta = \frac{V_{OUT} (V_{IN} (\alpha + 2) - V_{OUT} (\alpha + \beta + 4))}{V_{IN}^2 (\alpha + 1) - V_{IN} V_{OUT} (\alpha + 2) + \dfrac{16 K_{SW} P_{OUT} V_{OUT}}{A_c C_{den} (V_{IN} (2 + \alpha) - V_{OUT} (4 + \alpha + \beta))}} \tag{2.38}$$

Considering that the converter's area is mainly defined by the capacitors' area, then (2.38) allows to determine the converter's efficiency as a function of the power density ($P_{OUT}/Area$) for a given V_{IN}, V_{OUT}, and K_{SW}. For example, let's consider four different cases, where the transistors are all implemented by either 1.2 V NMOS transistors, 1.2 V PMOS transistors, 3.3 V NMOS transistors, or 3.3 V PMOS transistors.

Figure 2.16a and b show the efficiency (2.38) as a function of power density for $V_{SW} = V_{OUT} = 0.9$ V, different switches implementations, and for C_{FLY} implemented by a PMOS transistor ($C_{den} = 10$ fF/μm^2, $\alpha = 3\%$, and $\beta = 0\%$). The graphs show that 1.2 V

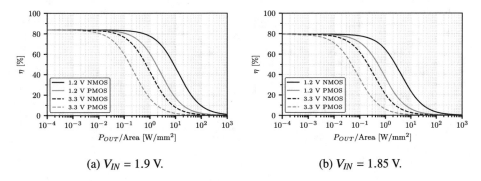

(a) $V_{IN} = 1.9$ V.   (b) $V_{IN} = 1.85$ V.

Fig. 2.16 Converter's efficiency as a function of P_{OUT}/Area, for different switches' implementation, with $V_{OUT} = V_{SW} = 0.9$ V, $C_{den} = 10$ fF, $\alpha = 3\%$, and $\beta = 0\%$

transistors are preferable in comparison with 3.3 V transistors. Furthermore, 1.2 V NMOS transistors allow to maximize the efficiency and the power density. However, the switch's implementation with 1.2 V transistors may not be feasible if the voltage across the transistor exceeds its breakdown voltage. Furthermore, unless one of the switch's terminals is close to ground, 1.2 V NMOS transistors may not be easy to implement. For example, the switch S_1 connects V_{IN} to the C_{FLY} top plate in ϕ_1. This means that the switch has to pass V_{IN}, requiring a gate voltage swing of at least 0.9 V higher than V_{IN}, increasing the driver's design complexity. On the other hand, a 1.2 V PMOS transistor would not require a voltage higher than V_{IN}, thus facilitating the driver's design.

Figure 2.17a and b show the efficiency (2.38) as a function of power density with $S_{1,2,4}$ implemented by PMOS transistors and S_3 by an NMOS transistor, C_{FLY} implemented by a PMOS transistor ($C_{den} = 10$ fF/μm^2, $\alpha = 3\%$, and $\beta \approx 0\%$), and for $V_{OUT} = V_{SW} = 0.9$ V. The graphs show that for V_{IN} values close to $V_{OUT}/CR = 1.8$ V, the maximum power density, whilst keeping efficiency constant, is within the 10–100 mW range, depending on either 1.2 V or 3.3 V transistors are used. The previous analysis allows to determine what is the maximum acceptable power density, whilst keeping the efficiency high, for a given technology.

Figure 2.18 shows the impact of both the parasitic capacitances from C_{FLY} ($\alpha = 3\%$ and $\beta = 0\%$), and the switches, on the converter's efficiency as a function of V_{IN}, for $V_{OUT} = V_{SW} = 0.9$ V, and for different power density values. The dashed line shows the ideal efficiency without any parasitics and the dash and dot line shows the efficiency only considering the impact of C_{FLY} parasitic capacitances. The solid lines show the efficiency for power densities values of 10, 100, and 1000 mW/mm^2 considering both the C_{FLY} and switch parasitic capacitances. Through this graph, it is possible to obtain the maximum achievable efficiency and it is possible to select the input voltage range in which the converters should work.

It is important to notice that once the switch is sized, the switches' R_{ON} is fixed throughout the whole V_{IN} range. In the previous graphs, the switch's R_{ON} was modified according to the V_{IN} value. In a real circuit, it is necessary to fix the minimum V_{IN} value for which the

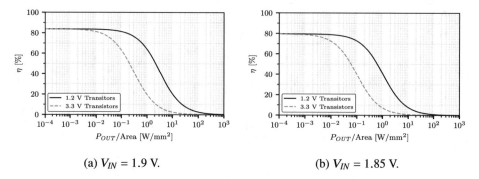

(a) $V_{IN} = 1.9$ V.   (b) $V_{IN} = 1.85$ V.

Fig. 2.17 Efficiency as a function of P_{OUT}/Area, with $S_{1,2,4}$ implemented by PMOS transistors and S_3 by an NMOS transistor, $V_{OUT} = V_{SW} = 0.9$ V, $C_{den} = 10$ fF, $\alpha = 3\%$, and $\beta = 0\%$

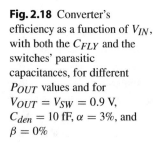

Fig. 2.18 Converter's efficiency as a function of V_{IN}, with both the C_{FLY} and the switches' parasitic capacitances, for different P_{OUT} values and for $V_{OUT} = V_{SW} = 0.9$ V, $C_{den} = 10$ fF, $\alpha = 3\%$, and $\beta = 0\%$

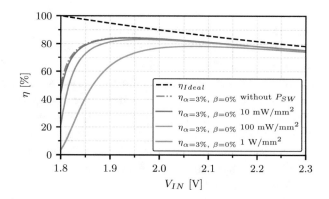

converter has to operate, therefore determining R_{ON}. Hence, the previous analysis is used to choose the minimum V_{IN} value, and then the efficiency is recalculated, using the calculated R_{ON} throughout the whole input range.

Figure 2.19a and b show an example where the solid lines are the efficiency recalculated using the R_{ON} fixed for the V_{IN} values of 1.8, 1.85, 1.9, 2.0 V, for 10 mW/mm² and 100 mW/mm², with $V_{OUT} = V_{SW} = 0.9$ V, $C_{den} = 10$ fF, $\alpha = 3\%$, and $\beta = 0\%$. The efficiency values after $V_{IN_{lim}}$ are not drawn because F_{CLK} increases beyond four time constants ($N = 4$). Hence, the circuit operates under very incomplete settling, making the previous equations no longer valid. These graphs show that as $V_{IN_{Lim}}$ gets close to V_{OUT}/CR there is a significant impact on the efficiency, especially at high power density values, e.g. 100 mW/mm². Thus, avoiding working close to V_{OUT}/CR is recommended because the value of R_{ON} is significantly low. Furthermore, the frequency increases rapidly close to the CR voltage value, hence any deviation from the ideal frequency value would cause the output voltage to quickly deviate from the 0.9 V target. Nonetheless, as the plots show, the previous analysis with the variable R_{ON} and with fixed R_{ON} are quite similar when working under the maximum power density (below 100 mW/mm²) and far from the V_{OUT}/CR value

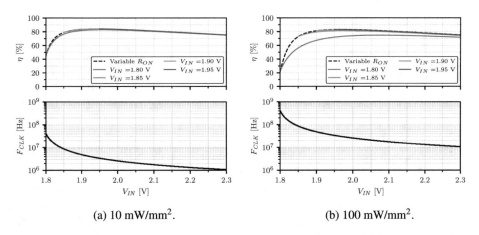

Fig. 2.19 Converter's efficiency as a function of V_{IN}, with both the C_{FLY} and the switches' parasitic capacitances, for a fixed R_{ON} value determined by the minimum V_{IN} value of the converter's operation. For $V_{OUT} = V_{SW} = 0.9$ V, $C_{den} = 10$ fF, $\alpha = 3\%$, and $\beta = 0\%$

($V_{IN} > 1.85$). The efficiency plot should be analysed together with F_{CLK} (2.18). For example, Fig. 2.19b, shows that to achieve a power density of 100 mW/mm² the converter must work at frequencies of 10–100 MHz, which adds complexity to the system design, mainly to the clock generator and the switch drivers. Hence, lower power densities, such as the ones in Fig. 2.19b, may be preferable to work at a lower F_{CLK} value.

2.4   Basic Voltage Regulation Technique

In the previous section, it was assumed that the output voltage, V_{OUT}, was kept constant. In reality, this voltage varies with each clock cycle due to the charge received from the flying capacitor, C_{FLY}. This voltage variation at the output is called voltage ripple, ΔV_{ripple}, which is undesirable because it results in extra power dissipation, and can compromise the performance of the circuits powered by the converter.

To better understand what is the output voltage ripple, lets look at the 1/1 SC converter (Fig. 2.20), where C_{FLY} charges to V_{IN} in phase ϕ_1 and it discharges in parallel with C_L and R_L, in phase ϕ_2, where C_L is the output decoupling capacitor and R_L the load resistance. Lets for now assume that $R_L \to \infty$. Then at the end of ϕ_1, C_{FLY} is charged to V_{IN} and C_L is discharged to V_{OUT}. At the beginning of ϕ_2, C_{FLY} and C_L are placed in parallel, and because $V_{IN} > V_{OUT}$, there is a voltage ripple ΔV_{ripple}. This value can be determined by

$$V_{IN} C_{FLY} + V_{OUT} C_L = (C_{FLY} + C_L)(V_{OUT} + \Delta V_{ripple}) \tag{2.39}$$

$$\Delta V_{ripple} = \frac{C_{FLY}}{C_{FLY} + C_L}(V_{IN} - V_{OUT}) \approx \frac{C_{FLY}}{C_L}(V_{IN} - V_{OUT}) \tag{2.40}$$

Fig. 2.20 Simplified schematic of a 1/1 SC converter, where the NMOS have their bulk connected to ground

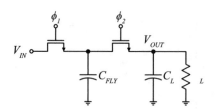

Equation (2.40) shows that to minimize ΔV_{ripple}, the ratio of C_{FLY}/C_L must be minimized. This can be made by using a large decoupling capacitor $C_L >> C_{FLY}$, which can be implemented on-chip. However, it results in a significant increase in the converter's area, due to the low capacitance density of on-chip capacitors. An external capacitor can also be used, nonetheless, it will increase the Bill of Materials (BOM) cost and requires an extra pad that might not be available. Another way to minimize ΔV_{ripple} is to regulate the amount of charge transferred to C_L [20].

Considering now, that $R_L \neq \infty$, the converter must be designed for the minimum V_{IN} value, which corresponds to the point of higher clock frequency ($F_{CLK_{MAX}}$) for a desired V_{OUT} and $P_{OUT_{MAX}}$ (or maximum output current I_L). Hence, with both the C_{FLY} and the switches' R_{ON} sized, the amount of charge transferred to the output node is the one required at $V_{IN_{MIN}}$ at $P_{OUT_{MAX}}$. When either P_{OUT} or V_{IN} are not equal to their maximum values, F_{CLK} must be adjusted to control the amount of charge transferred to the output, and thus keeping V_{Ripple} small (2.41) [21]. Hence, the most common way to control the charge sent to the output is to use frequency modulation.

$$\Delta V_{Ripple} = \frac{I_L}{2\,C_{FLY}\,F_{CLK}} \tag{2.41}$$

Frequency modulation can be implemented using a hysteretic control. It was first introduced in [22] and it was composed of two voltage boundaries that would start or stop the clock phases of the converter, to keep V_{OUT} within the boundaries. Therefore, two comparators were required, which resulted in a slower response by the controller and could result in stability issues. This controller was later simplified by having a single comparator, that feeds or not the clock phases to the converter, this is called single boundary hysteretic control [23, 24]. Since there is only one comparator, hence one boundary, the control complexity decreases leading to higher speed and few stability issues.

Figure 2.21 shows the block diagram of the single boundary hysteretic controller. It features a comparator, an AND gate, and a clock generator, which can be implemented by an Asynchronous state machine (ASM) or by an oscillator. The working principle is quite simple, let V_{REF} be the desirable voltage reference for the converter to settle, then if $V_{REF} - V_{OUT} > 0$, the comparator V_C goes high, enabling the AND gate, and thus letting the clock signal (CLK) pass to the phase generator, which generates the clock phases ϕ_1 and ϕ_2 for the converter. This brings V_{OUT} back to the V_{REF} level. When $V_{REF} - V_{OUT} < 0$,

Fig. 2.21 Single bound hysteretic control block diagram for loop regulation

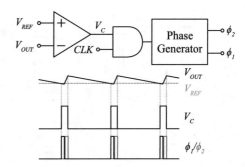

the clock signal is stopped due to $V_C = 0$, and V_{OUT} discharges at a rate determined by the output decoupling capacitor. Ideally, the voltage cross point for the comparator to shut down the clock generator would be when $V_{REF} - V_{OUT} = 0$. However, the comparator will have a non-zero time response delay and also a non-zero offset voltage. Hence, there will be a delay in the start and shutting down of the clock, which will result in output ripple voltage.

Since having a large output decoupling capacitor is undesirable, a technique called multiphase interleaving has been widely used to completely eliminate the need of an external decoupling capacitor [4, 10, 14, 20, 21, 24–34]. There are two ways to implement the interleaving technique, the first, shown in Fig. 2.22, it works by dividing the converter into N smaller cells, that are switched T_{CLK}/N delayed from each other, resulting in an evenly spaced charge transfer over the whole switching period. This reduces both the output voltage and input current ripple, because the charge taken from the input and delivered to the output is made in smaller amounts, at different time instants, instead of a large amount of charge at a single time instant. Equation (2.41) can now be re-written as [6, 21]

$$V_{Ripple} = \frac{I_L}{N \, C_{TOTAL} \, F_{SW}} \tag{2.42}$$

where C_{TOTAL} is the total flying capacitance plus any decoupling capacitance seen by the output. Again, the higher the number of interleaved cells, the lower the voltage ripple. However, higher N values lead to an increase in the design complexity of the phase generator and also an increase in the complexity of the routing of the clock signals to the converter cells.

The second way to implement the interleaving technique is to duplicate the converter cell and operate it in phase opposition [16, 17, 20, 34–36], as shown in Fig. 2.23. This technique is mostly used in converters that cascaded multiple cells, because it results in a lesser complex layout. However, the decrease in the voltage ripple is not as notorious as the N interleaved cells, because the cell is delivering the maximum charge in each phase, whilst when dividing the cells, the charge is divided by the number of cells.

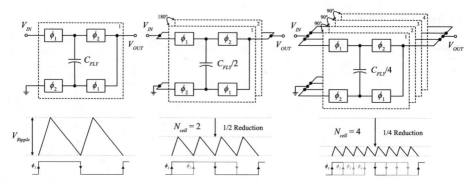

Fig. 2.22 N-cells time interleaved scheme, where $C_{TOTAL} = C_{FLY}$

Fig. 2.23 Converter with two
cells in phase opposition,
where $C_{TOTAL} = 2\,C_{FLY}$

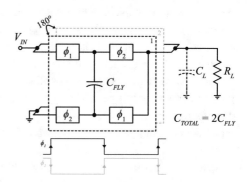

References

1. Razavi B (2011) Design of analog CMOS integrated circuits. McGraw-Hill, New York
2. Carusone TC, Johns D, Martin K (2011) Analog integrated circuit design. Wiley, Hoboken
3. Breussegem TV, Steyaert M (2013) CMOS integrated capacitive DC-DC converters. Springer, New York
4. Jiang J, Lu Y, Huang C, Ki W, Mok PKT (2015) A 2-/3-phase fully integrated switched-capacitor DC-DC converter in bulk CMOS for energy-efficient digital circuits with 14% efficiency improvement. In: 2015 IEEE international solid-state circuits conference (ISSCC). https://doi.org/10.1109/ISSCC.2015.7063078
5. Ramadass YK, Chandrakasan AP (2007) Voltage scalable switched capacitor dc-dc converter for ultra-low-power on-chip applications. In: IEEE power electronics specialists conference. https://doi.org/10.1109/PESC.2007.4342378
6. Bang S, Seo J, Chang L, Blaauw D, Sylvester D (2016) A low ripple switched-capacitor voltage regulator using flying capacitance dithering. IEEE J Solid-State Circuits. https://doi.org/10.1109/JSSC.2015.2507361
7. Butzen N, Steyaert M (2016) Scalable parasitic charge redistribution: design of high-efficiency fully integrated switched-capacitor DC-DC converters. IEEE J Solid-State Circuits. https://doi.org/10.1109/JSSC.2016.2608349

8. Baker RJ (2010) CMOS circuit design, layout, and simulation. Wiley-IEEE Press

9. Meyvaert H, Sarafianos A, Butzen N, Steyaert M (2014) Monolithic switched-capacitor DC-DC towards high voltage conversion ratios. In: IEEE 15th workshop on control and modeling for power electronics (COMPEL). https://doi.org/10.1109/COMPEL.2014.6877191

10. El-Damak D, Bandyopadhyay S, Chandrakasan AP (2013) A 93% efficiency reconfigurable switched-capacitor DC-DC converter using on-chip ferroelectric capacitors. In: IEEE international solid-state circuits conference digest of technical papers. https://doi.org/10.1109/ISSCC.2013.6487776

11. Chang L, Montoye RK, Ji BL, Weger AJ, Stawiasz KG, Dennard RH (2010) A fully-integrated switched-capacitor 2:1 voltage converter with regulation capability and 90% efficiency at 2.3A/mm^2. In: 2010 symposium on VLSI circuits. https://doi.org/10.1109/VLSIC.2010.5560267

12. Andersen TM, Krismer F, Kolar JW, Toifl T, Menolfi C, Kull L, Morf T, Kossel M, Brändli M, Buchmann P, Francese PA (2013) A 4.6 W/mm^2 power density 86% efficiency on-chip switched capacitor DC-DC converter in 32 nm SOI CMOS. In: 2013 twenty-eighth annual IEEE applied power electronics conference and exposition (APEC). https://doi.org/10.1109/APEC.2013.6520285

13. Andersen TM, Krismer F, Kolar JW, Toifl T, Menolfi C, Kull L, Morf T, Kossel M, Brändli M, Buchmann P, Francese PA (2014) A deep trench capacitor based 2:1 and 3:2 reconfigurable on-chip switched capacitor DC-DC converter in 32 nm SOI CMOS. In: 2014 IEEE applied power electronics conference and exposition (APEC). https://doi.org/10.1109/APEC.2014.6803497

14. Andersen TM, Krismer F, Kolar JW, Toifl T, Menolfi C, Kull L, Morf T, Kossel M, Brändli M, Buchmann P, Francese PA (2015) A feedforward controlled on-chip switched-capacitor voltage regulator delivering 10W in 32nm SOI CMOS. In: IEEE international solid-state circuits conference (ISSCC). https://doi.org/10.1109/ISSCC.2015.7063076

15. Andersen TM, Krismer F, Kolar JW, Toifl T, Menolfi C, Kull L, Morf T, Kossel M, Brändli M, Buchmann P, Francese PA (2017) A 10 W on-chip switched capacitor voltage regulator with feedforward regulation capability for granular microprocessor power delivery. IEEE Trans Power Electron. https://doi.org/10.1109/TPEL.2016.2530745

16. Bang S, Wang A, Giridhar B, Blaauw D, Sylvester D (2013) A fully integrated successive-approximation switched-capacitor DC-DC converter with 31mV output voltage resolution. In: IEEE international solid-state circuits conference (ISSCC). https://doi.org/10.1109/ISSCC.2013.6487774

17. Bang S, Blaauw D, Sylvester D (2016) A successive-approximation switched-capacitor DC–DC converter with resolution of $V_{IN}/2^N$ for a wide range of input and output voltages. https://doi.org/10.1109/JSSC.2015.2501985

18. Harjani R, Chaubey S (2014) A unified framework for capacitive series-parallel DC-DC converter design. In: Proceedings of the IEEE 2014 custom integrated circuits conference. https://doi.org/10.1109/CICC.2014.6946050

19. Seeman MD, Sanders SR (2008) Analysis and optimization of switched-capacitor DC-DC converters. IEEE Trans Power Electron. https://doi.org/10.1109/TPEL.2007.915182

20. Kudva SS, Harjani R (2013) Fully integrated capacitive DC-DC converter with all-digital ripple mitigation technique. IEEE J Solid-State Circuits. https://doi.org/10.1109/JSSC.2013.2259044

21. Le H, Sanders SR, Alon E (2011) Design techniques for fully integrated switched-capacitor DC-DC converters. IEEE J Solid-State Circuits. https://doi.org/10.1109/JSSC.2011.2159054

22. Seeman MD, Sanders SR, Rabaey JM (2008) An ultra-low-power power management IC for energy-scavenged wireless sensor nodes. In: IEEE power electronics specialists conference. https://doi.org/10.1109/PESC.2008.4592048

23. Seeman M, Jain R (2011) Single-bound hysteretic regulation of switched capacitor converters. Intel Corp. https://patents.google.com/patent/US20110074371A1/en
24. Breussegem TMV, Steyaert MSJ (2011) Monolithic capacitive DC-DC converter with single boundary-multiphase control and voltage domain stacking in 90 nm CMOS. IEEE J Solid-State Circuits. https://doi.org/10.1109/JSSC.2011.2144350
25. Le H, Seeman M, Sanders SR, Sathe V, Naffziger S, Alon E (2010) A 32nm fully integrated reconfigurable switched-capacitor DC-DC converter delivering 0.55W/mm^2 at 81% efficiency. In: IEEE international solid-state circuits conference - (ISSCC). https://doi.org/10.1109/ISSCC.2010.5433981
26. Piqué GV (2021) A 41-phase switched-capacitor power converter with 3.8mV output ripple and 81% efficiency in baseline 90nm CMOS. In: IEEE international solid-state circuits conference. https://doi.org/10.1109/ISSCC.2012.6176892
27. Le H, Crossley J, Sanders SR, Alon E (2013) A sub-ns response fully integrated battery-connected switched-capacitor voltage regulator delivering 0.19W/mm^2 at 73% efficiency. In: IEEE international solid-state circuits conference (ISSCC). https://doi.org/10.1109/ISSCC.2013.6487775
28. Jain R, Geuskens B, Khellah M, Kim S, Kulkarni J, Tschanz J, De V (2013) A 0.45–1V fully integrated reconfigurable switched capacitor step-down DC-DC converter with high density MIM capacitor in 22nm tri-gate CMOS. In: IEEE symposium on VLSI circuits
29. Andersen TM, Krismer F, Kolar JW, Toifl T, Menolfi C, Kull L, Morf T, Kossel M, Brändli M, Buchmann P, Francese PA (2014). IEEE international solid-state circuits conference digest of technical papers (ISSCC). https://doi.org/10.1109/ISSCC.2014.6757351
30. Sarafianos A, Steyaert M (2015) Fully integrated wide input voltage range capacitive DC-DC converters: the folding dickson converter. IEEE J Solid-State Circuits. https://doi.org/10.1109/JSSC.2015.2410800
31. Biswas A, Sinangil Y, Chandrakasan AP (2015) A 28 nm FDSOI integrated reconfigurable switched-capacitor based step-up DC-DC converter with 88% peak efficiency. IEEE J Solid-State Circuits. https://doi.org/10.1109/JSSC.2015.2416315
32. Tsai J, Ko S, Wang C, Yen Y, Wang H, Huang P, Lan P, Shen M (2015) A 1 V input, 3 V-to-6 V output, 58%-efficient integrated charge pump with a hybrid topology for area reduction and an improved efficiency by using parasitics. IEEE J Solid-State Circuits. https://doi.org/10.1109/JSSC.2015.2465853
33. Lu Y, Jiang J, Ki W (2017) A Multiphase Switched-Capacitor DC-DC Converter Ring With Fast Transient Response and Small Ripple. IEEE J Solid-State Circuits. https://doi.org/10.1109/JSSC.2016.2617315
34. Jiang Y, Law M, Chen Z, Mak P, Martins RP (2019) Algebraic series-parallel-based switched-capacitor DC-DC boost converter with wide input voltage range and enhanced power density. IEEE J Solid-State Circuits. https://doi.org/10.1109/JSSC.2019.2935556
35. Salem LG, Mercier PP (2014) An 85%-efficiency fully integrated 15-ratio recursive switched-capacitor DC-DC converter with 0.1-to-2.2V output voltage range. In: IEEE international solid-state circuits conference (ISSCC). https://doi.org/10.1109/ISSCC.2014.6757350
36. Lutz D, Renz P, Wicht B (2016) A 10mW fully integrated 2-to-13V-input buck-boost SC converter with 81.5% peak efficiency. In: IEEE international solid-state circuits conference (ISSCC). https://doi.org/10.1109/ISSCC.2016.7417988

Fully Integrated Switched-Capacitor DC-DC State of the Art

3

In the previous chapter, the behaviour and performance analysis of Switched Capacitor (SC) DC-DC converters were described in detail. It was concluded that the choice of the passive devices to implement the capacitor and switches has a strong impact on the converter's overall performance. Hence, the technological node chosen to implement the converter will dictate its performance. Furthermore, the Series-Parallel (SP) 1/2 step-down converter was also analysed in detail in the previous chapter, however, there are several other topologies that can implement different ratios, and thus, a multi-ratio SC converter can be made by combining different topologies. Hence, this chapter presents a comprehensive literature review of the current trends of fully integrated multi-ratio SC converters.

3.1 Topologies for Multi-ratio Fully Integrated SC DC-DC Converters

As shown in the previous chapter, the maximum efficiency of an SC converter is achieved close to its Conversion Ratio (CR). Hence, if the converter has to generate a wide output voltage range, or convert a wide input voltage range into a stable output voltage, then different CRs have to be combined, to maximize the efficiency throughout the whole voltage range. Considering that separated converters would result in a large area, the multi-ratio converters are achieved by the recombination of the switches and capacitors. The next sections will describe the most used topologies in multi-ratio converters.

3.1.1 Series-Parallel Topology

SP converters use one or more flying capacitors (C_{FLY}) either in series or in parallel, but never a combination of both, to achieve a certain CR [1]. Any CR can be achieved, depending on the number of capacitors. As an example, Fig. 3.1 shows the schematics of two SP SC converter topologies 1/2 and 2/3, where the first, 1/2, has been already explained in detail in

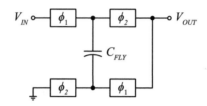

(a) Schematic of the 1/2 topology.

(b) Schematic of the 1/2 topology in ϕ_1.   (c) Schematic of the 1/2 topology in ϕ_2.

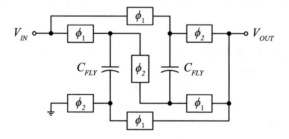

(d) Schematic of the 2/3 topology.

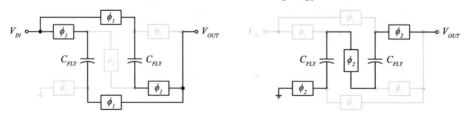

(e) Schematic of the 2/3 topology in ϕ_1.   (f) Schematic of the 2/3 topology in ϕ_2.

Fig. 3.1 Simplified schematic of the 1/2 and 2/3 SP SC converter topologies [1]

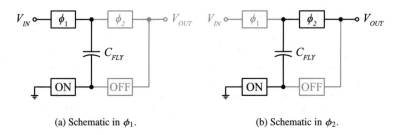

(a) Schematic in ϕ_1.   (b) Schematic in ϕ_2.

Fig. 3.2 Simplified schematic implementation of the SP 1/1 SC converter out of the SP 1/2 topology schematic

the previous chapter. The 2/3 topology can be easily understood, during phase ϕ_1, both C_{FLY} capacitors charge to $V_{IN} - V_{OUT}$, giving $V_{C_{FLY}} = V_{IN} - V_{OUT}$. During phase ϕ_2, both C_{FLY} discharge in series with V_{OUT}, hence $V_{OUT} = 2\,V_{C_{FLY}}$. Combining both equations gives that $V_{OUT} = 2/3\,V_{IN}$. This circuit topology has 1 more C_{FLY} and 3 more switches than the circuit that implements the 1/2 converter.

Multiple ratios can be achieved by combining different SP topologies. Where by controlling the switches clock phase, the flying capacitors can be configured into different configurations. For example, Fig. 3.2 shows the same 1/2 topology, as in Fig. 3.1a, but now configured to implement a 1/1 CR. Due to its simplicity multi-ratio SP SC converters have gain interest in the past few years [2–13].

Table 3.1 shows a summary of the comparison of fully integrated multi-ratio SP SC converters, implemented in bulk Complementary Metal-Oxide-Semiconductor (CMOS) technology. Considering the step-down topologies [2, 3, 5, 7, 8, 14], high efficiency (>80%) is achieved. However, due to the reduced number of CRs ($\leq$5), the average efficiency within the voltage range is significantly lower than the maximum value. Consequently, the circuit's voltage range is typically below 1 V, except for the works [5, 8], which have the largest conversion input voltage range value of 2.3 V and 1.6 V, respectively. Although in [5] the 1/2 CR is used to cover an input voltage range of almost 1.3 V, which results in a very low average efficiency ($\approx$40%). This reduced number of CRs is due to the layout complexity increase with the increase in the number of CRs, and also the losses due to the higher number of switches and capacitors. Furthermore, in the majority of the works, the output power values, and consequently power density values, are in the range of tenths of milliwatts, except for the works presented in [7, 14], which reported power density values of 180 and 164 mW/mm^2, respectively. Nonetheless, in [7], the peak efficiency occurs at 66.6 mW/mm^2, after that, the efficiency rapidly decreases and can be as low as 60%. In [14] the reported power density value is 2.15 W/mm^2, which does not match with the area and power values reported. The power density value calculated through the converter's area, and the maximum power, is 164 mW/mm^2, and similar to [7] it is not maintained through all the CRs. The works [5, 6, 8] do not use phase interleaved, and thus, have lower power density values. Considering the step-up topologies [5, 15], these typically have lower maximum efficiency and respective

Table 3.1 Comparison table summary of multi-ratio SC converters using SP topology

Publication	[2]'12	[3]'13	[5]'15	[7]'17	[14]'17	[8]'18	[15]'20
Tech. (nm)	90	130	65	65	65	180	180
CR #	2	2	3	3	5	4	3
CR	Down: 1/2, 2/3	Down: 1/2, 1/3	Down 1/2, 2/3, 1/1 Up: 3/2, 2/1, 3/1	Down: 1/2, 2/3, 3/4	Down: 1/3, 1/2, 2/3 3/4, 1/1	Down: 2/3, 3/5, 1/2, 2/5	Up: 2, 3/2, 4/3
Switch #	9	8	N.A.	12	N.A.	14	12
C_{FLY} (Type)	Thin-gate MOS, MOM	MIM	N.R.	MOS, MOM, MIM	MOS, MOM, MIM	MIM	MOS, MIM
C_{OUT} (Type)	Thin-gate MOS, MOM	MOS, MIM	N.R.	MOS, MOM, MIM	MOS, MOM, MIM	MOS	MOS
C_{OUT} (nF)	0.084	5	N.R.	2	0.45	5	0.2
Phase interleaved	Yes (41)	Yes (2)	No	Yes (123)	Yes (6)	No	Yes (101)
V_{IN} (V)	1.2–2	1.2	0.5–3.3	1.6–2.2	1.0	2–3.6	2.6–4.2
V_{IN} range (V)	0.8	–	2.5	0.6	–	1.6	1.6
V_{OUT} (V)	0.7	0.3–0.55	1	0.6–1.2	0.25–0.95	1.2	3.5–4.3
V_{OUT} range (V)	–	0.25	–	0.6	0.7	–	0.8
V_{Ripple} (mV)	3.8	50	N.R.	30	30	90	52
Area (mm^2)	0.25	0.97	0.48	0.84	1.24[a]	1.06	3.47
η_{max} (%)	81	70	Down: 69[a] Up: 70.4	80	82	84.2	82
P_{OUT} at η_{max} (mW)	5.6	24.5	0.0034	56	203	2.4	258
$P_{density}$ at η_{max} (mW/mm^2)	38.6	3.94[a]	0.007[a]	66.6	164.04[a]	2.26	54.5
$P_{density_{max}}$ (mW/mm^2)	38.6	24.5	0.007[a]	180	164.04[a]	2.40	54.5

[a]Estimated from the corresponding literature, N.A. not applicable, N.R. not reported

Table 3.2 Comparison table summary of multi-ratio SC converters using SP topology, with exotic capacitors to implement C_{FLY}

Publication	[9]'11	[10]'13	[11]'14	[12]'15
Tech. (nm)	32 SOI	22 Tri-gate	32 SOI	32 SOI
CR #	3	4	2	2
CR	Down: 1/2, 2/3, 1/3	Down: 1/2, 2/3, 4/5 1/1	Down: 1/2, 2/3	Down: 1/2, 2/3
Switch #	9	10	12	8
C_{FLY} (Type)	Thin-Oxide MOS	HD-MIM	Deep Trench	Deep Trench
C_{OUT} (Type)	None	None	None	None
C_{OUT} (nF)	0	0	0	0
Phase interleaved	Yes (32)	Yes (8)	Yes (16)	Yes (64)
V_{IN} (V)	2	1.23	1.8	1.8
V_{IN} range (V)	–	–	–	
V_{OUT} (V)	0.4–1.1	0.45–1.05	0.7–1.1	0.7–1.1
V_{OUT} range (V)	0.7	0.6	0.4	0.4
V_{Ripple} (mV)	N.R.	130	30	30
Area (mm^2)	0.378	0.101	0.15	1.968
η_{max} (%)	79.76	84.2	90	85.1
P_{OUT} at η_{max} (mW)	325	13	840	10×10^3
$P_{density}$ at η_{max} (mW/mm^2)	860	128.24[a]	3.71×10^3	3.2×10^3
$P_{density_{max}}$ (mW/mm^2)	1.7×10^{3a}	355.11[a]	5.7×10^{3a}	5.1×10^{3a}

[a]Estimated from the corresponding literature, N.A. not applicable, N.R. not reported

low power density values when compared with the step-down ones. However, in [15] has reported a peak efficiency value of 82% and a power density of 54.5 mW/mm^2, which is similar to the ones found in step-down converters.

Table 3.2 shows the comparison of step-down fully integrated multi-ratio SP SC converter implemented with high density on-chip capacitors [9–12]. Using this type of capacitor, high power density can be achieved. However, these capacitors are not part of bulk CMOS technology, and therefore require additional masks, which have additional costs. Finally, most of the aforementioned works are not completely fully integrated, in the sense that some do not have load regulation [2, 9], or do not have a CR controller [10, 11], or do not have a clock generator or reference voltage generator [3, 5, 7, 8, 12, 14] on-chip.

3.1.2 Dickson Topology

The Dickson converter was initially proposed in 1976 [16], shown in Fig. 3.3, and has also been receiving attention in recent years [17–20]. In the recent Integrated Circuit (IC) versions, the major change is the replacement of the diodes by CMOS switches to eliminate the diode voltage drop [19]. The Dickson converter can be used to step-up [21–23] or to step-down [19, 20, 24] the input voltage. And since it is a regular structure, it allows for a modular and compact design, and for its stages to be folded into each other, easily creating several conversion ratios while maximizing the use of the flying capacitors [19]. Figure 3.4 shows the conventional step-down 1/N Dickson SC converter. Where by alternating the phases of each stage a 1/N CR can be achieved by having $N - 1$ stages. As in the SP topology, the Dickson topology can also be used to generate multiple ratios. Figure 3.5 shows a two-stage converter that can implement the ratios 1/2 and 1/3 depending on how the switches are configured. For example, Fig. 3.5a and b show the converter schematic in both phases to implement the CR of 1/3. In phase ϕ_1, V_{C1} is equal to $V_{IN} - V_{OUT}$ and V_{C2} to V_{OUT}. This results in a V_{OUT}, in ϕ_2 of

$$V_{OUT} = V_{C1} + V_{C2} = V_{IN} - 2\,V_{OUT} \Rightarrow V_{OUT} = \frac{1}{3}\,V_{IN} \tag{3.1}$$

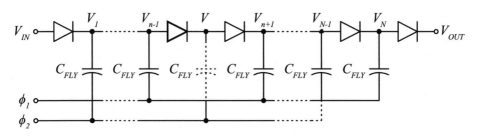

Fig. 3.3 Simplified schematic of the Dickson SC converter voltage multiplier [16]

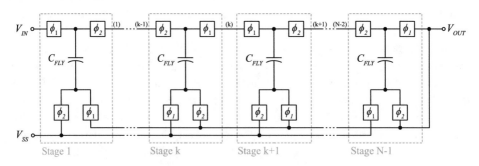

Fig. 3.4 Simplified schematic of a step-down 1/N Dickson SC converter [24]

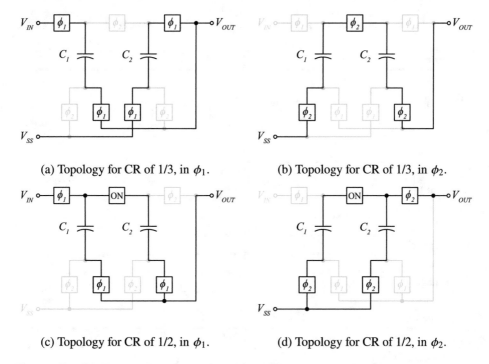

(a) Topology for CR of 1/3, in ϕ_1.   (b) Topology for CR of 1/3, in ϕ_2.

(c) Topology for CR of 1/2, in ϕ_1.   (d) Topology for CR of 1/2, in ϕ_2.

Fig. 3.5 Simplified schematic of a two-stage Dickson SC converter topology [1]

Figure 3.5c and d show the schematic of the same converter now used to implement the CR of 1/2. In phase ϕ_1, V_{C1} and V_{C2} are equal, and are given by $V_{IN} - V_{OUT}$. Hence, this results in a V_{OUT}, in ϕ_2 of

$$V_{OUT} = V_{C1} = V_{C2} = V_{IN} - V_{OUT} \Rightarrow V_{OUT} = \frac{1}{2} V_{IN} \qquad (3.2)$$

In [19, 20], multi-ratio converters using the Dickson topology are presented. However, only the work [19] has on-chip flying capacitances. Table 3.3 shows the performance of a multi-ratio step-down Dickson converter. It achieves a power density of 16.3 mW/mm², similar to the SP topologies power density values. However, it covers a much wider input voltage range (5.2 V) with a peak efficiency of 76.6%. As in the SP converters, the average efficiency is lower than the peak efficiency. Also, as the number of stages increases so do the losses increase, and the converter's start-up is inefficient, since all the capacitors have to be charged from ground [25]. This work is not completely fully integrated, in the sense that it does not have an on-chip load regulation, nor a CR controller, nor a clock generator, nor a reference voltage generator.

Table 3.3 Performance summary of a step-down multi-ratio SC converter using the Dickson topology

Publication	[19]'15
Tech. (nm)	90
CR #	4
CR	Down: 1/2, 1/3, 1/4, 1/5
Switch #	13
C_{FLY} (Type)	MIM
C_{OUT} (Type)	None
C_{OUT} (nF)	0
Phase interleaved	Yes (9)
V_{IN} (V)	2.8–8
V_{IN} range (V)	5.2
V_{OUT} (V)	1.2
V_{OUT} range (V)	–
V_{Ripple} (mV)	N.R.
Area (mm^2)	3.07
η_{max} (%)	76.6
P_{OUT} at η_{max} (mW)	32
$P_{density}$ at η_{max} (mW/mm^2)	10.42
$P_{density_{max}}$ (mW/mm^2)	16.3

N.R. not reported

3.1.3　Customized Topologies

There are other topologies, like Ladder [26, 27], Fibonacci [28, 29], Symmetric Cockcroft-Walton (SCW) [23, 30, 31], or Doubler [32, 33] topologies. However, most of them are either not fully integrated or have a single CR. Nonetheless, there are works that implement multi-ratio fully integrated converters, that do not fit on the previous two topologies or are a combination of them [23, 25, 34–42].

In the previous section, the Dickson 1/3 CR (Fig. 3.5) was able to generate a 1/3 CR by subtract V_{C_2} voltage by V_{C1} voltage. This subtraction property was used in combination with SP topologies in [37, 38, 42] to overcame the Dickson's $1/N$, or $\times N$, CR limitation, where in [42], a topology named Algebraic Series-Parallel (ASP) was used to implement a rational step-up SC converter. A rational boost CR can be defined as

$$CR = \frac{V_{OUT}}{V_{IN}} = K + \frac{m}{n} \qquad (3.3)$$

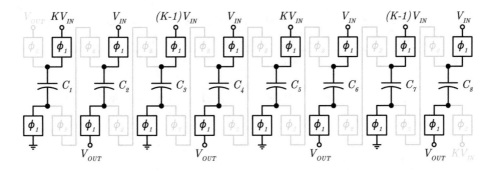

(a) Schematic in ϕ_1.

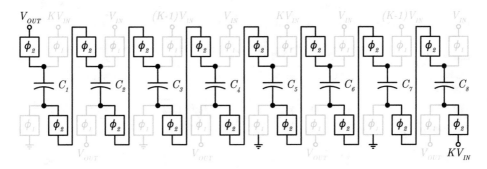

(b) Schematic in ϕ_2.

Fig. 3.6 Simplified schematic of the ASP based topology with $m/n = 2/5$ [42]

where $K, m, n \in N^+$ with $m \leq n$ and m, n are relatively prime. To achieve this ratio, in ϕ_1 the capacitors are charged in parallel either to $V_{IN} - V_{OUT}$, if the stage number (i) is even, or to $(K - p_i) V_{IN}$, if the stage number (i) is odd, where p_i is a configuration factor, given by

$$p_i = \begin{cases} 1, & \left(\frac{i+1}{2}\right)\left(1 - \frac{m}{n}\right) - \sum_{K=0}^{(i-1)/2} p_{2K-1} > 1 \\ 0, & \left(\frac{i+1}{2}\right)\left(1 - \frac{m}{n}\right) - \sum_{K=0}^{(i-1)/2} p_{2K-1} < 1 \end{cases} \tag{3.4}$$

This assures that the voltage swing on the capacitors is minimized, to minimize the charge redistribution and parasitic capacitance losses. On phase ϕ_2, the capacitors are all in series with V_{IN}, and thus V_{OUT} is the sum of all the capacitor voltages plus V_{IN}. Hence, the CR equation (3.3), can rewritten as

$$CR = K + \frac{N_F - 2\sum_{K=1}^{N_F/2} p_{2K-1}}{N_F + 2} \tag{3.5}$$

where N_F is the number of flying capacitors.

Table 3.4 Comparison table summary of multi-ratio SC converters using customized topologies

Publication	[35]'13	[37]'15	[23]'15	[39]'16	[27]'17	[42]'19	[41]'20
Tech. (nm)	65	65	180	180	130	65	65
CR #	2	2	2	2	5	7	4
CR	Down: 1/3, 2/5	Down: 1/4, 1/3	Up: 4/1, 6/1	Up: 2/1, 3/2	Down: 1/4, 1/3, 1/2, 2/3, 1/1	Up: 5/4, 3/2, 3/5, 2/1, 5/2, 3/1, 5/1	Down: 1/3, 2/3, 3/4, 4/5
Switch #	15	8	20	8	16	25	16
C_{FLY} (Type)	MOS	MOS, MOM MIM	MOS, MIM	N.R.	MOS	MOS, MIM	MOS, MOM MIM
C_{OUT} (Type)	None	N.R.	MOS, MIM	N.R.	MOS	N.R.	N.R.
C_{OUT} (nF)	–	N.R.	0.054	4.05	0.76	N.R.	N.R.
Phase interleaved	Yes (18)	Yes (2)	Yes (9)	No	No	Yes (2)	N.R.
V_{IN} (V)	3–4	1.5–2.5	1	0.35–0.6	2.2	0.25–1	N.R
V_{IN} range (V)	1	1	–	0.25	–	0.75	N.R
V_{OUT} (V)	1	0.4–0.7	3–6	0.86–1.8	0.2–1.1	1	N.R.
V_{OUT} range (V)	–	0.3	3	0.94	0.9	–	N.R.
V_{Ripple} (mV)	N.R.	N.R.	40	N.R	N.R.	84	N.R.
Area (mm^2)	0.64	0.23	0.5	1.75	0.291	0.54	0.3
η_{max} (%)	74.3	79.5	58	75.8	80.6[b]	80[b]	85
P_{OUT} at η_{max} (mW)	122	26	1.44	0.4	1.97[a]	22.7[b]	0.6[a]
$P_{density}$ at η_{max} (mW/mm^2)	40[a]	33.6[a]	2.50[a]	0.23	6.68[a,b]	22.7[b]	3.2[a]
$P_{density_{max}}$ (mW/mm^2)	190	113.4[a]	2.88[a]	N.R.	7.56[a,b]	37.8[a,b]	24[b]

[a]Estimated from the corresponding literature, N.R. not reported
[b]External load regulation

Figure 3.6 shows an example of the ASP topology used to implement a ratio m/n of 2/5, with $p_{1,3,5,7} = \{0\ 1\ 0\ 1\}$. For example, with $K = 1$, it results in a CR of 7/5, and with $K = 2$ it results in a CR of 12/5. However, $K > 1$ requires K voltage levels for the conversion. Which have to be generated, for example, by other converters. Hence, this topology was implemented for $K = 1$ in [42], and was combined with the SP voltage doubler and Dickson topologies.

Table 3.4 shows the comparison of multi-ratio SC converter that use customized topologies and are implemented in bulk CMOS technology. Considering the step-down topologies [27, 35, 37, 41], the numbers of voltage range, CRs, efficiency, and power density values, are close to the ones of the previous topologies. As for the step-up converters [23, 39, 42], efficiency and power density values have significantly increase when compared with the previous topologies. Especially in [42], which achieves good efficiency and power density values for a converter in bulk CMOS. However, the converters cannot achieve the same output power level for all ratios. For example, the ratios 3/2 and 2, achieve the highest output power, 20.4 mW, the ratios 5/4 and 3/5 achieve ≈14 mW, the 5/2 and 3 achieve ≈10 mW, and finally, ratio 5 only achieves ≈5 mW. Hence only 25% of the maximum output power level is covered by all ratios. Moreover, in terms of efficiency, it rapidly decreases to ≈65%, in the ratios 3 and 5/2, and to ≈47%, in the 5 ratio, resulting in a difference of 30% less than the peak efficiency. Finally, no on-chip load regulation control, or CR controller, or start-up circuit, was implemented in this system. Which would also decrease both the efficiency and power density.

Finally, only the work in [39] is fully integrated. The other works, similarly to the previous topologies, are not completely fully integrated in the sense that some do not have on-chip load regulation [27, 41, 42], or do not have a CR controller [41, 42], or do not have a clock generator or reference voltage generator [23, 27, 35, 37, 41, 42].

3.1.4 Cascaded Topologies

One way to increase the CRs number, the voltage range, and the average efficiency, is to connect several SC converters in series (Fig. 3.7). However, the overall efficiency decreases with the number of stages ($_{tot} = _1 \times _2 \times \cdots \times _N$) and this also results in an increase of the output impedance (decreases the maximum output current capacity) [43]. Moreover, the load impedance seen by the first stages is very high, meaning that these stages are working very close to their CR voltage limit. Hence, when considering the parasitic capacitors, even 1% is enough to drop the efficiency significantly, at values close to the CR. This is a problem especially for lighter power values and prevents the use of higher capacitance per area devices (e.g. MOS capacitors) because the parasitics are higher and thus efficiency will be lower.

Furthermore, when cascading converters, a decoupling capacitor (C_{DC}) between stages is required (Fig. 3.8a). One way to eliminate the decoupling capacitor is to place another 1/2 converter in parallel using the complementary clock phases (Fig. 3.8b) [44, 45]. However, both solutions require extra capacitors and switches that will decrease the power density per area [46]. The need of extra capacitors can be eliminated by operating the next stage with twice the clock frequency than the previous stage [46], thus saving area. However, now the clock frequency increases exponentially with the number of stages $F_{CLK_{MAX}} = 2^N F_{CLK_{\#1}}$.

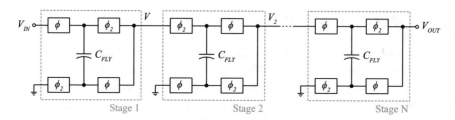

Fig. 3.7 Simplified schematic of N SP 1/2 SC converter in series

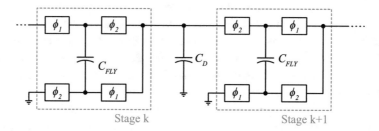

(a) Inter-stage decoupling capacitor C_D.

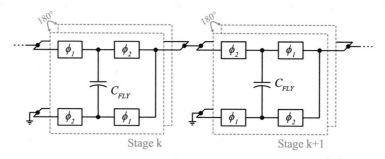

(b) Parallel 1/2 cell on opposite phase.

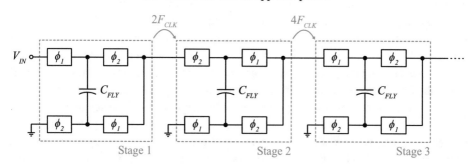

(c) Doubling F_{CLK} between stages.

Fig. 3.8 Inter-stage decoupling capacitor solutions when cascading 1/2 SC converters [46]

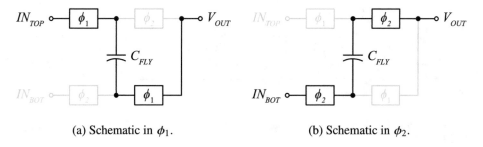

(a) Schematic in ϕ_1.   (b) Schematic in ϕ_2.

Fig. 3.9 Simplified schematic of the dual-input 1/2 SC converter [44]

This means that either the overall converter's frequency is reduced, thus decreasing the power density, or the control circuits will have to work at very high frequencies, resulting in larger power dissipation and increasing the design complexity.

One can look at the 1/2 SP SC converter as a system with two inputs a one output, as is shown in Fig. 3.9. The two inputs, IN_{TOP} and IN_{BOT} result in an output voltage given by

$$V_{OUT} = \frac{1}{2} (IN_{TOP} + IN_{BOT}) \tag{3.6}$$

The converter, named dual-input 1/2 converter, can now be used to sum two input voltages. This widens the number of CRs that can be achieved and now the efficiency is no longer the stage's efficiency multiplication, and thus, high efficiency can be achieved. This is shown in the works [44, 45, 47–49], were several 1/2 converters, named cells, were connected in series (cascaded) to achieve several CR. Notice that the same is true for the 2/1 converter, where in this case in ϕ_1, C_{FLY} is charge to IN_{TOP}, and on ϕ_2, it is placed in series with IN_{BOT} and V_{OUT}, giving $V_{OUT} = IN_{TOP} + IN_{BOT}$ [48].

In [44, 45], a Successive-Approximation-Register (SAR) SC converter is presented. The stages are fed either by $(V_{HIGH}, V_{MID_{n-1}})$ or $(V_{MID_{n-1}}, V_{LOW})$. Figure 3.10 shows an examples where the 4 cells (4 bits) are cascaded. Multiplexers are used to select the IN_{TOP} (or IN_{BOT}) that propagates through the multiplexers and the last output from the previous stage. The only exception being the first cell, that is connected to V_{IN} and V_{SS}. In this figure, an output voltage of 1.125 V is produced out of 2 V by using the digital word 1000 (a_3, a_2, a_1, a_0). The theoretical resolution is given by $V_{OUT} = V_{IN}/2^N$, where N is the number of stages, resulting in a resolution of 0.125 V.

The works [46, 47, 49, 50] also present a Recursive SC (RSC) converter topology similar to the previous work. The difference is that now the cells are binary weighted and V_{IN}/V_{SS} can now be connected to any stage, unlike the previous work, where if V_{MID} was used for the next stage IN_{TOP}, it would become the highest possible voltage for the next stages.

Figure 3.11 shows an example of the dual-input 1/2 binary-reconfigurable SC converter for obtaining a CR of 3/8, i.e. $V_{TARGET} = 0.75V$ from $V_{IN} = 2V$ developed in [47, 50]. The first cell has always its IN_{TOP} and IN_{BOT} connected to V_{IN} and V_{SS}, respectively, which

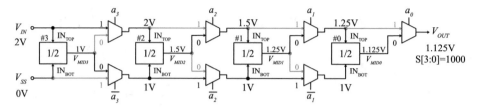

Fig. 3.10 Simplified structure of the 4 bit SAR SC converter with the code 1000, resulting in $V_{OUT} = 1.125$ V [44, 45]

Fig. 3.11 Simplified schematic of the binary-reconfigurable SC converter [47, 50]

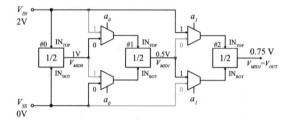

generates an output voltage $V_{MID0} = 1$ V. The multiplexer's select the signals according with the following rules: if $V_{MID} > V_{TARGET}$, then $a = 0$, and so $IN_{TOP} = V_{MID}$ and $IN_{BOT} = V_{SS}$; if $V_{MID} < V_{TARGET}$, then $a = 1$, and $IN_{TOP} = V_{IN}$ and $IN_{BOT} = V_{MID}$. This repeats until the output voltage of the stage V_{MID_N} is equal to V_{TARGET}. This technique minimizes the amount of charge through the flying capacitors, thus maximizing the capacitance utilization and minimizing cascaded losses. The number of iterations defines the resolution of $2^N - 1$. Moreover, because the CR is 1/2 from stage to stage, the first stages have an exponentially smaller impact than the ones close to the output, thus the cells are binary weighted. This can be seen through the following equation:

$$V_{OUT} = V_N = \frac{1}{2}\left(V_{IN} \cdot a_n + V_{MID-1}\right) = \tag{3.7}$$

$$= \frac{1}{2}\left(V_{IN} \cdot a_N + \frac{1}{2}\left(V_{IN} \cdot a_{N-1} + V_{MID-2}\right)\right) = \tag{3.8}$$

$$= V_{IN}\left(\frac{a_N}{2} + \frac{a_{N-1}}{4} + \frac{a_{N-2}}{8} + \cdots + \frac{a_1}{2^N}\right) \tag{3.9}$$

where a_n (0 or 1) defines if it is V_{IN} that is fed to the stage ($IN_{TOP} = V_{IN}$ and $IN_{BOT} = V_{MID_n}$), or if its V_{MID} ($IN_{TOP} = V_{MID_n}$ and $IN_{BOT} = V_{SS}$). Although the authors of this work increased the power density by more than 35 times, when compared to [44, 45], this value is still far from the values of the few CRs converters [5, 7, 19, 35, 42].

The previous works only fed to the cells inputs V_{IN}, V_{MID}, or V_{SS}, and thus the CRs were limited to 2^N. However, more CRs can be generated by introducing more voltage values in the stages inputs. This idea was presented in [51], where the converter can achieve any

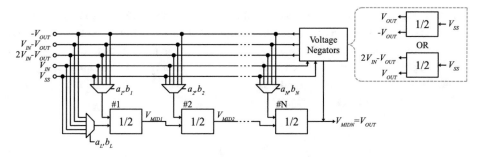

Fig. 3.12 Structure of the Rational-CR SC converter Using Negative-Output Feedback [51]

arbitrary binary ratio: $p/q, 0 < p < q \leq 2^{N+1}$, where N is the number of cascaded stages. The novelty is to incorporate a negative voltage feedback into the cascaded converter stages by using negative-generating converter sates (called *Voltage Negators*) [52]. With this, both the numerator p and the denominator q can be reconfigured. This technique allows to achieve output voltage conduction values close to the ones found in few discrete ratios, like in the series-parallel topologies.

Figure 3.12 shows the structure of the Rational-CR SC converter Using Negative-Output Feedback. As in [47, 50], the output of the previous 1/2 cell is connected to one of the inputs of the following stage. However, now the other stage' input can connect to either V_{IN}, V_{SS}, $-V_{OUT}$, $V_{IN} - V_{OUT}$, and $2V_{IN} - V_{OUT}$. Where V_{OUT} is given by

$$V_{OUT} = A \cdot V_{IN} - B \cdot V_{OUT} = V_{IN} \cdot \frac{A}{1+B} \qquad (3.10)$$

where A and B are the converter' forward path gain and feedback factor, respectively. The negative voltage feedbacks provide three extra choices for each stage, increasing the number of CRs. Furthermore, the negating converters give extra current to the output, thus improving the overall output conductance.

To determine the Multiplexer (MUX) configuration both A and B are determined by

$$A = \frac{p}{2^N} \qquad (3.11)$$

$$B = \frac{A \cdot q - p}{p} \qquad (3.12)$$

where N is a natural number that verifies $0 < p < q \leq 2^{N+1}$. By converting both A and B to binary fractional numbers, $A_{(2)} = [a_1, a_2, \ldots, a_N]_{(2)}$ and $B_{(2)} = [b_1, b_2, \ldots, b_N]_{(2)}$, allows for the MUXs configuration through the following conditions:

$$MUX_i = \begin{cases} a_i \cdot V_{IN} - b_i \cdot V_{OUT}, & \text{if } A < 1 \\ a_i \cdot V_{IN} - b_i(V_{OUT} - V_{DD}), & \text{if } A \geq 1 \end{cases} \qquad (3.13)$$

This means that the $-V_{OUT}$ line is only used when $A < 1$ and, the $2V_{IN} - V_{OUT}$ line when $A \geq 1$. The rational converter can provide more CR than the previous binary converters. Furthermore, the number of CRs increases faster with the number of cascaded stages than when using binary converters. To better illustrate how it is configured, an example is depicted below. For a CR of 4/13, the required number of cells is 3 ($2^{3+1} = 16$), and the converter's forward gain and feedback factors are

$$A = \frac{p}{2^N} = \frac{4}{2^3} = 0.5 = 0.100_{(2)} \tag{3.14}$$

$$B = \frac{A \cdot q - p}{p} = \frac{0.5 \times 13 - 4}{4} = 0.625 = 0.101_{(2)} \tag{3.15}$$

Since $A = 0.5 < 1$, the MUXs will be configured as $MUX_i = a_i \cdot V_{IN} - b_i \cdot V_{OUT}$ (3.13), and Fig. 3.9a shows the configuration of the converter for obtaining the 4/13 CR.

As another example, let's look at the 9/11 CR configuration:

$$A = \frac{p}{2^N} = \frac{9}{2^3} = 1.125 = 1.001_{(2)} \tag{3.16}$$

$$B = \frac{A \cdot q - p}{p} = \frac{1.125 \times 11 - 9}{9} = 0.375 = 0.011_{(2)} \tag{3.17}$$

$$A' = A - B = 0.110_{(2)} \tag{3.18}$$

Since $A = 1.125 > 1$, then its replaced by $A' = A - B$, and the MUXs will be configured as $MUX_i = V_{IN} - b_i(V_{OUT} - V_{DD})$ (3.13), and Fig. 3.9b shows the configuration of the converter for obtaining the 4/13 CR (Fig. 3.13).

The previous works all used low parasitic Metal-Insulator-Metal (MIM) capacitors to achieve high efficiency, which limits their power density to <1 mW/mm^2. In [53], a fully integrated fine-grained buck-boost SC converter with 24 CR was proposed. It uses an Algorithmic Voltage-Feed-In (AVFI) topology to systematically generate any arbitrary buck/boost rational ratio so that it minimizes both the conduction loss and the parasitic losses. The converter is divided into ten main SC cell (MC) and other ten auxiliary SC cell (AC) (1/5 of the size of the MC), that have two configurations: Dickson or charge-path folding (QF). Each rational cell (RC) branch is fed by either V_{IN} or V_{OUT}, depending on the desired CR. It can provide any CR from $(n + 1) : 1$ to $(n + 1) : n$ in the buck mode and from $n : (n + 1)$ to $1 : (n + 1)$ in boost mode, respectively.

Figure 3.14 shows the eight possible configurations of the each RC stage. The Voltage-feed-in (VFI) coefficients (a_i and b_i) determine how V_{IN} and V_{OUT} are connect in each cell and m_i determines the configuration either Dickson of QF, according to the desired CR. Concerning the parasitic capacitances, the voltage across them should be minimized. In this work, the bottom parasitic capacitance is dominant, and thus ΔV_{CP} should be minimized. The Dickson configuration intrinsically ensures charger transfer from the top plates, and bottom plates, of the current stage to the next ($C_{i,top} \longrightarrow C_{i+1,top}$ and $C_{i,bot} \longrightarrow C_{i+1,bot}$),

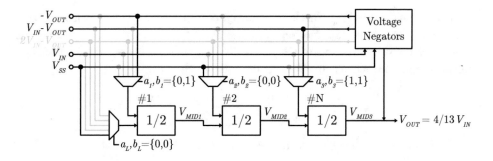

(a) CR of 4/13.

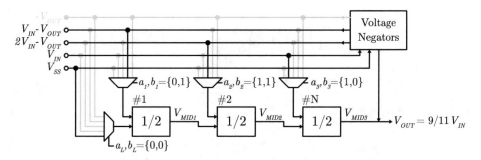

(b) CR of 9/11.

Fig. 3.13 Example of the Rational-CR SC converter configurations [51]

m_i	Buck (b_i=1)		Boost (a_i=1)		m_i
a_i	0	1	0	1	a_i
0	V_i V_{i+1} V_{SS} V_{OUT} TT (Dickson)	V_i V_{OUT} V_{SS} V_{i+1} TB (QF)	V_i V_{i+1} V_{IN} V_{SS} TT (Dickson)	V_{IN} V_{i+1} V_i V_{SS} BT (QF)	0
1	V_{IN} V_{OUT} V_i V_{i+1} BB (Dickson)	V_{IN} V_{i+1} V_i V_{OUT} TB (QF)	V_{IN} V_{OUT} V_i V_{i+1} BB (Dickson)	V_i V_{OUT} V_{IN} V_{i+1} TB (QF)	1

Fig. 3.14 RC configuration summary [53]

and thus it minimizes ΔV_{CB}. This is an advantage of the Dickson topology when compared to the binary (including RSC and Negator-based SC (NSC)) and series-parallel structures, which have transitions patterns of $C_{i,top} \longrightarrow C_{i+1,bot}$ and $C_{i,bot} \longrightarrow C_{i+1,top}$, leading to higher ΔV_{CP}. The system parameters for the buck are given by

$$a_1 = 0, \quad a_{i \in \{N^+ | 2 \le i \le n\}} = \begin{cases} 1, & \text{if } 1 + \sum_{j=1}^{i-1} a_j < i \cdot CR \\ 0, & \text{if } 1 + \sum_{j=1}^{i-1} a_j > i \cdot CR \end{cases} \tag{3.19}$$

$$b_{i \in \{N^+ | i \le n\}} = 1, \quad m_{i \in \{N^+ | i \le n\}} = a_i \oplus a_{i+1}, \quad m_n = X \text{ (either 0 or 1)} \tag{3.20}$$

and for the boost, by

$$a_{i \in \{N^+ | i \le n\}} = 1 \tag{3.21}$$

$$b_{i \in \{N^+ | i \le n-1\}} = \begin{cases} 1, & \text{if } 1 + \sum_{j=1}^{i-1} b_j < (i+1)/CR \\ 0, & \text{if } 1 + \sum_{j=1}^{i-1} b_j > (i+1)/CR \end{cases}, \quad b_n = 0, \tag{3.22}$$

$$m_1 = X, \text{ (either 0 or 1)}, \quad m_{i \in \{N^+ | 2 \le i \le n\}} = b_{i-1} \oplus b_i \tag{3.23}$$

with the CR determined by

$$CR = \frac{q}{p} = \frac{1 + \sum_{i=1}^{n} a_i}{1 + \sum_{i=1}^{n} b_i} \quad \text{with} \quad \begin{cases} p = n + 1, \ 1 \le q \le n \\ q = n + 1, \ 1 \le p \le n \end{cases} \tag{3.24}$$

Figure 3.15 shows an example of the implementation of the CR 4/7. Using the equations above, the parameters are given by $a_{1-6} = 010101_{(2)}$, $b_{1-6} = 1$, and $m_{1-6} = 11111X_{(2)}$.

The authors claim that the quasi-Dickson property allows for a reduction of approximately 50% of the parasitic-loss-factor ($M_{PAR} = \sum [a_{ci}(\Delta V_{CB}/V_{IN})^2]$) in the buck mode and, in the boost mode, the M_{PAR} has a quasi-linear parasitic loss profile, instead of the exponentially increase as in RSC and NSC. The efficiency has an increase of approximately 6% in most CRs, when compared to the previously described topologies.

Table 3.5 shows a performance comparison of the aforementioned converters. When compared with the non-cascaded topologies, the input/output voltage range was greatly increased

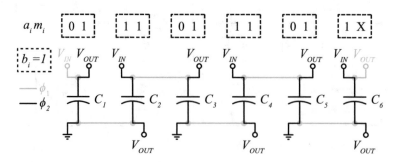

Fig. 3.15 Example of the 4/7 Buck AVFI topology configuration [53]

Table 3.5 Comparison table summary of multi-ratio SC converters using cascaded topologies

Publication	[44]'13	[50]'14	[49]'16	[51]'16	[48]'16	[53]'18
Tech. (nm)	180	250	350 HV	180	180	65
CR #	20 (117)	15	17	79	14	24
Type	Step-down	Step-down	Step-down/up	Step-down	Step-up	Step-down/up
C_{FLY} (Type)	MIM	MIM	MIM	MIM	HD-MIM	MOS, MIM
C_{OUT} (Type)	None	None	MIM	None	HD-MIM	MOS
C_{OUT} (nF)	–	–	N.R.	–	2.05	6nF
V_{IN} (V)	3.4–4.3	2.5	2.0–13.0	2	0.45–3	0.22–2.4
V_{IN} range (V)	0.9	–	11	–	2.55	2.18
V_{OUT} (V)	0.9–1.5	0.1–2.2	5	0.47–1.87	3.3	0.85–1.2
V_{OUT} range (V)	0.6	2.1	–	1.4	–	0.35
V_{Ripple} (mV)	81	N.R.	N.R.	N.R.	N.R.	Down: 90 Up: 51
Area (mm^2)	1.69	4.645	6.8	2.436	4	2.4
η_{max} (%)	72	85	Down: 81.5 Up: 70.9	95[b]	81	Down: 84.1[c] Up: 83.2[c]
$P_{density}$ at η_{max} (mW/mm^2)	0.009[a]	4.4[a]	Down: 0.96[a] Up: 0.15[a]	0.13[a,b]	0.013[a]	Down: 13.4 Up: 10.8
$P_{density_{max}}$ (mW/mm^c)	0.45[a]	4.4[a]	Down: 1.47[a] Up: 1.02[a]	0.13[a,b]	0.013[a]	Down: 34.17 Up: 14.17[a]

[a]Estimated from the corresponding literature
[b]External regulation
[c]Power from external sources such as external clock and voltage reference are not included

due to the increase in the number of CRs and it has resulted in an increase in the average efficiency. However, it is still significantly lower than the maximum efficiency value. For example, in [53] the boost ratios reach minimum efficiency values of 20–30% and buck ratio values of 60–70%. These are quite far from the 84.1% peak efficiency value. Furthermore, the power density values are lower than the non-cascaded topologies, and thus, the total area is much larger, approximately 6 times larger, than the average area of the SP converters. Therefore, increasing the design complexity of these converters.

Finally, similarly to the previous topologies, these are not completely fully integrated, in the sense that some do not have load regulation [51], or do not have CR regulation [51, 53], or do not have a clock generator or reference voltage generator [44, 48–51, 53] on-chip.

3.2   Conclusions

Figure 3.16 shows the State-of-the-Art (SoA) graph of multi-ratio SC converters with on-chip flying capacitors implemented in CMOS technology. The graph shows the maximum efficiency (η_{max}) versus the respective power density, with the converters grouped by the technology used to implement the flying capacitors. Except for the outstanding efficiency value of [51], high efficiency and power density are only achieved when using exotic capacitors, like ferroelectric or deep trench capacitors in Silicon-On-Insulator (SOI) technology. This is because the capacitance density is larger, and the parasitic capacitances are smaller than the ones in bulk CMOS technology.

Since one of the advantages of fully integrating the Power Management Unit (PMU) is the cost reduction, it makes sense to look at the SoA considering only the bulk CMOS technology, which the cost is much lower than the SOI technologies or the ones using an extra mask for the exotic capacitors. Figure 3.17 show the multi-ratio SC converters' with on-chip capacitances SoA considering only the converters implemented in bulk CMOS technology. Now, the maximum power density is below 200 mW/mm². Moreover, except for the work [51], the maximum efficiency is around 85% and decreases as the power density increases. Figure 3.17a shows the SoA grouped by the technology used to implement the

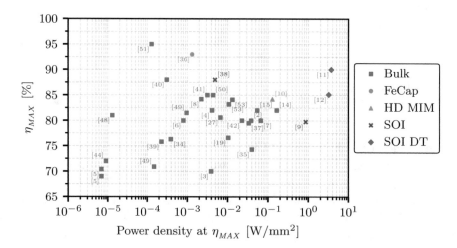

Fig. 3.16 State of the Art of multi-ratio SC converters with on-chip capacitors: maximum efficiency versus the respective power density

converter. The most commonly used technology is 65 nm and is the one that achieves the best trade-off between the efficiency and the power density. As explained in the previous chapter, the smaller the technology of the node, the smaller the oxide thickness. Hence, the capacitance per area increases, and thus it is expected that, as the technology node decreases, the power density increases. Figure 3.17b shows the SoA is grouped by topologies. The graph shows that cascaded topologies achieve the highest efficiency, while discrete topologies (SP, Dickson, and others) achieve the highest power density. Finally, Fig. 3.17c shows the SoA is grouped by the step-down and step-up topologies. Overall, the performance of step-up converters is worse than the step-down converter. Only recent works [15, 42, 53], published in

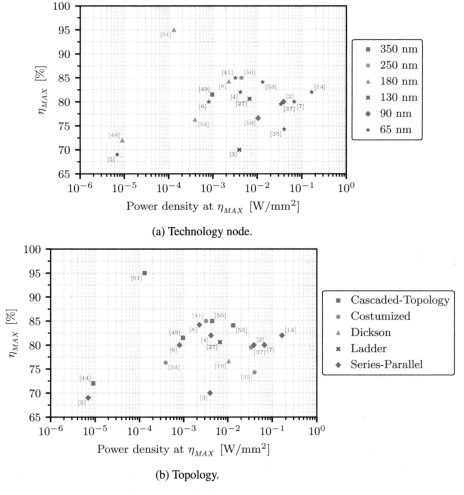

(a) Technology node.

(b) Topology.

Fig. 3.17 State of the Art of multi-ratio SC converters with on-chip capacitors and considering only bulk CMOS technologies: maximum efficiency versus respective power density

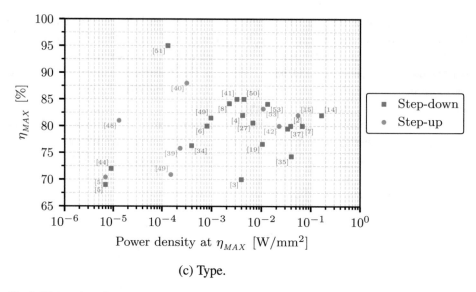

(c) Type.

Fig. 3.17 (continued)

2018, 2019, and 2020, respectively, reach a performance similar to the step-down converters. However, works [42, 53] do not have on-chip load regulation, nor CR-regulation, nor clock and voltage reference generator.

References

1. Harjani R, Chaubey S (2014) A unified framework for capacitive series-parallel DC-DC converter design. In: Proceedings of the IEEE 2014 custom integrated circuits conference. https://doi.org/10.1109/CICC.2014.6946050
2. Piqué GV (2021) A 41-phase switched-capacitor power converter with 3.8mV output ripple and 81% efficiency in baseline 90nm CMOS. In: IEEE international solid-state circuits conference. https://doi.org/10.1109/ISSCC.2012.6176892
3. Kudva SS, Harjani R (2013) Fully integrated capacitive DC-DC converter with all-digital ripple mitigation technique. IEEE J Solid-State Circuits. https://doi.org/10.1109/JSSC.2013.2259044
4. Myers J, Savanth A, Howard D, Gaddh R, Prabhat P, Flynn D (2015) An 80 nW retention 11.7 pJ/cycle active subthreshold ARM Cortex-M0+ subsystem in 65 nm CMOS for WSN applications. In: 2015 IEEE international solid-state circuits conference - (ISSCC) digest of technical papers. https://doi.org/10.1109/ISSCC.2015.7062967
5. Hua X, Harjani R (2015) 3.5–0.5 V input, 1.0 V output multi-mode power transformer for a supercapacitor power source with a peak efficiency of 70.4%. In: 2015 IEEE custom integrated circuits conference (CICC). https://doi.org/10.1109/CICC.2015.7338390
6. Kilani D, Alhawari M, Mohammad B, Saleh H, Ismail M (2016) An efficient switched-capacitor DC-DC buck converter for self-powered wearable electronics. IEEE Trans Circuits Syst I Regul Pap. https://doi.org/10.1109/TCSI.2016.2586117

7. Lu Y, Jiang J, Ki W (2017) A multiphase switched-capacitor DC-DC converter ring with fast transient response and small ripple. IEEE J Solid-State Circuits. https://doi.org/10.1109/JSSC.2016.2617315

8. ianxi L, Hao C, Tianyuan H, Junchao M, Zhangming Z, Yintang Y (2018) A dual mode step-down switched-capacitor DC-DC converter with adaptive switch width modulation. Microelectr J https://doi.org/10.1016/j.mejo.2018.06.003

9. Le H, Sanders SR, Alon E (2011) Design techniques for fully integrated switched-capacitor DC-DC converters. IEEE J Solid-State Circuits. https://doi.org/10.1109/JSSC.2011.2159054

10. Jain R, Geuskens B, Khellah M, Kim S, Kulkarni J, Tschanz J, De V (2013) A 0.45–1V fully integrated reconfigurable switched capacitor step-down DC-DC converter with high density MIM capacitor in 22nm tri-gate CMOS. In: IEEE symposium on VLSI circuits

11. Andersen TM, Krismer F, Kolar JW, Toifl T, Menolfi C, Kull L, Morf T, Kossel M, Brändli M, Buchmann P, Francese PA (2014) IEEE international solid-state circuits conference digest of technical papers (ISSCC). https://doi.org/10.1109/ISSCC.2014.6757351

12. Andersen TM, Krismer F, Kolar JW, Toifl T, Menolfi C, Kull L, Morf T, Kossel M, Brändli M, Buchmann P, Francese PA (2015) A feedforward controlled on-chip switched-capacitor voltage regulator delivering 10W in 32nm SOI CMOS. In: IEEE international solid-state circuits conference (ISSCC). https://doi.org/10.1109/ISSCC.2015.7063076

13. Andersen TM, Krismer F, Kolar JW, Toifl T, Menolfi C, Kull L, Morf T, Kossel M, Brändli M, Buchmann P, Francese PA (2017) A 10 W On-Chip Switched Capacitor Voltage Regulator With Feedforward Regulation Capability for Granular Microprocessor Power Delivery. IEEE Trans Power Electron. https://doi.org/10.1109/TPEL.2016.2530745

14. Saurabh C, Harjani R (2017) Fully tunable software defined DC-DC converter with 3000X output current 4X output voltage ranges. In: IEEE custom integrated circuits conference (CICC). https://doi.org/10.1109/CICC.2017.7993625

15. Junmin J, Xun L, Wing-Hung K, Philip MKT, Yan L (2020) A multiphase switched-capacitor converter for fully integrated AMLED microdisplay system. IEEE Trans Power Electron. https://doi.org/10.1109/TPEL.2019.2951799

16. Dickson JF (1976) On-chip high-voltage generation in MNOS integrated circuits using an improved voltage multiplier technique. IEEE J Solid-State Circuits. https://doi.org/10.1109/JSSC.1976.1050739

17. Cao D, Peng FZ (2010) A family of zero current switching switched-capacitor DC-DC converters. In: IEEE applied power electronics conference and exposition (APEC). https://doi.org/10.1109/APEC.2010.5433407

18. Cao D, Jiang S, Peng FZ (2013) Optimal design of a multilevel modular capacitor-clamped DC-DC converter. IEEE Trans Power Electron. https://doi.org/10.1109/TPEL.2012.2231438

19. Sarafianos A, Steyaert M (2015) Fully integrated wide input voltage range capacitive DC-DC converters: the folding dickson converter. IEEE J Solid-State Circuits. https://doi.org/10.1109/JSSC.2015.2410800

20. Ng V, Sanders S (2012) A 92%-efficiency wide-input-voltage-range switched-capacitor DC-DC converter. In: IEEE international solid-state circuits conference. https://doi.org/10.1109/ISSCC.2012.6177016

21. Wang Y, Yan N, Min H, Shi CJR (2017) A high-efficiency split-merge charge pump for solar energy harvesting. IEEE Trans Circuits Syst II Express Briefs. https://doi.org/10.1109/TCSII.2016.2581589

22. Shih YC, Otis BP (2011) An inductorless DC-DC converter for energy harvesting with a 1.2-μW bandgap-referenced output controller. IEEE Trans Circuits Syst II: Express Briefs. https://doi.org/10.1109/TCSII.2011.2173967

23. Tsai J, Ko S, Wang C, Yen Y, Wang H, Huang P, Lan P, Shen M (2015) A 1 V input, 3 V-to-6 V output, 58%-efficient integrated charge pump with a hybrid topology for area reduction and an improved efficiency by using parasitics. IEEE J Solid-State Circuits. https://doi.org/10.1109/JSSC.2015.2465853
24. Butzen N, Steyaert MSJ (2017) Design of soft-charging switched-capacitor DC-DC converters using stage outphasing and multiphase soft-charging. IEEE J Solid-State Circuits. https://doi.org/10.1109/JSSC.2017.2733539
25. Wu X, Shi Y, Jeloka S, Yang K, Lee I, Lee Y, Sylvester D, Blaauw D (2017) A 20-pW discontinuous switched-capacitor energy harvester for smart sensor applications. IEEE J Solid-State Circuits. https://doi.org/10.1109/JSSC.2016.2645741
26. Steyaert M, Tavernier F, Meyvaert H, Sarafianos A, Butzen N (2015) When hardware is free, power is expensive! Is integrated power management the solution? In: European solid-state circuits conference (ESSCIRC). https://doi.org/10.1109/ESSCIRC.2015.7313820
27. Castro Lisboa P, Pérez-Nicoli P, Veirano F, Silveira F (2016) General top/bottom-plate charge recycling technique for integrated switched capacitor DC-DC converters. IEEE Trans Circuits Syst I Regul Pap. https://doi.org/10.1109/TCSI.2016.2528478
28. Cabrini A, Gobbi L, Torelli G (2007) Voltage gain analysis of integrated fibonacci-like charge pumps for low power applications. IEEE Trans Circuits Syst II Express Briefs. https://doi.org/10.1109/TCSII.2007.904156
29. Makowski MS, Maksimovic D (1995) Performance limits of switched-capacitor DC-DC converters. In: Proceedings of PESC '95 - power electronics specialist conference. https://doi.org/10.1109/PESC.1995.474969
30. Zhang R, Huang Z, Inoue Y (2009) A low breakdown-voltage charge pump based on Cockcroft-Walton structure. In: IEEE 8th international conference on ASIC. https://doi.org/10.1109/ASICON.2009.5351433
31. Tsai J, Tseng C, Tseng W, Shia TK, Huang P (2012) An integrated 12-V electret earphone driver with symmetric Cockcroft-Walton pumping topology for in-ear hearing aids. In: IEEE asian solid state circuits conference (A-SSCC). https://doi.org/10.1109/IPEC.2012.6522623
32. Pelliconi R, Iezzi D, Baroni A, Pasotti M, Rolandi PL (2003) Power efficient charge pump in deep submicron standard CMOS technology. IEEE J Solid-State Circuits. https://doi.org/10.1109/JSSC.2003.811991
33. TBreussegem TV, Steyaert M (2009) A 82% efficiency 0.5% ripple 16-phase fully integrated capacitive voltage doubler. In: 2009 symposium on VLSI circuits
34. Ramadass YK, Chandrakasan AP (2007) Voltage scalable switched capacitor DC-DC converter for ultra-low-power on-chip applications. In: IEEE power electronics specialists conference. https://doi.org/10.1109/PESC.2007.4342378
35. Le H, Crossley J, Sanders SR, Alon E (2013) A sub-ns response fully integrated battery-connected switched-capacitor voltage regulator delivering 0.19W/mm^2 at 73% efficiency. In: IEEE international solid-state circuits conference (ISSCC). https://doi.org/10.1109/ISSCC.2013.6487775
36. El-Damak D, Bandyopadhyay S, Chandrakasan AP (2013) A 93% efficiency reconfigurable switched-capacitor DC-DC converter using on-chip ferroelectric capacitors. In: IEEE international solid-state circuits conference digest of technical papers. https://doi.org/10.1109/ISSCC.2013.6487776
37. Jiang J, Lu Y, Huang C, Ki W, Mok PKT (2015) A 2-/3-phase fully integrated switched-capacitor DC-DC converter in bulk CMOS for energy-efficient digital circuits with 14% efficiency improvement. In: 2015 IEEE international solid-state circuits conference (ISSCC). https://doi.org/10.1109/ISSCC.2015.7063078

38. Biswas A, Sinangil Y, Chandrakasan AP (2015) A 28 nm FDSOI integrated reconfigurable switched-capacitor based step-up DC-DC converter with 88% peak efficiency. IEEE J Solid-State Circuits. https://doi.org/10.1109/JSSC.2015.2416315

39. Ozaki T, Hirose T, Asano H, Kuroki N, Numa M (2016) Fully-integrated high-conversion-ratio dual-output voltage boost converter with MPPT for low-voltage energy harvesting. IEEE J Solid-State Circuits. https://doi.org/10.1109/JSSC.2016.2582857

40. Rawy K, Yoo T, Kim TT (2018) An 88% efficiency 0.1–300-μ W energy harvesting system with 3-D MPPT using switch width modulation for IoT smart nodes. IEEE J Solid-State Circuits. https://doi.org/10.1109/JSSC.2018.2833278

41. Jiang J, Liu X, Huang C, Ki W, Mok PKT, Lu Y (2020) Subtraction-mode switched-capacitor converters with parasitic loss reduction. IEEE Trans Power Electron. https://doi.org/10.1109/TPEL.2019.2933623

42. Jiang Y, Law M, Chen Z, Mak P, Martins RP (2019) Algebraic series-parallel-based switched-capacitor DC-DC boost converter with wide input voltage range and enhanced power density. IEEE J Solid-State Circuits. https://doi.org/10.1109/JSSC.2019.2935556

43. Teh CK, Suzuki A (2016) A 2-output step-up/step-down switched-capacitor DC-DC converter with 95.8% peak efficiency and 0.85-to-3.6V input voltage range. In: IEEE international solid-state circuits conference (ISSCC). https://doi.org/10.1109/ISSCC.2016.7417987

44. Bang S, Wang A, Giridhar B, Blaauw D, Sylvester D (2013) A fully integrated successive-approximation switched-capacitor DC-DC converter with 31mV output voltage resolution. In: IEEE international solid-state circuits conference (ISSCC). https://doi.org/10.1109/ISSCC.2013.6487774

45. Bang S, Blaauw D, Sylvester D (2016) A successive-approximation switched-capacitor DC–DC converter with resolution of $V_{IN}/2^N$ for a wide range of input and output voltages. https://doi.org/10.1109/JSSC.2015.2501985

46. Salem LG, Mercier PP (2015) A battery-connected 24-ratio switched capacitor PMIC achieving 95.5%-efficiency. In: Symposium on VLSI circuits (VLSI Circuits). https://doi.org/10.1109/VLSIC.2015.7231315

47. Salem LG, Mercier PP (2014) An 85%-efficiency fully integrated 15-ratio recursive switched-capacitor DC-DC converter with 0.1-to-2.2V output voltage range. IEEE International Solid-State Circuits Conference (ISSCC), https://doi.org/10.1109/ISSCC.2014.6757350

48. Liu X, Huang L, Ravichandran K, Sánchez-Sinencio E (2016) A highly efficient reconfigurable charge pump energy harvester with wide harvesting range and two-dimensional MPPT for internet of things. IEEE J Solid-State Circuits. https://doi.org/10.1109/JSSC.2016.2525822

49. Lutz D, Renz P, Wicht B (2016) A 10mW fully integrated 2-to-13V-input buck-boost SC converter with 81.5% peak efficiency. In: IEEE international solid-state circuits conference (ISSCC). https://doi.org/10.1109/ISSCC.2016.7417988

50. Salem LG, Mercier PP (2014) A recursive switched-capacitor DC-DC converter achieving $2^N -$ 1 ratios with high efficiency over a wide output voltage range. IEEE J Solid-State Circuits. https://doi.org/10.1109/JSSC.2014.2353791

51. Jung W, Sylvester D, Blaauw D (2016) A rational-conversion-ratio switched-capacitor DC-DC converter using negative-output feedback. In: IEEE international solid-state circuits conference (ISSCC). https://doi.org/10.1109/ISSCC.2016.7417985

52. Jung W, Oh S, Bang S, Lee Y, Foo Z, Kim G, Zhang Y, Sylvester D, Blaauw D (2014) An ultra-low power fully integrated energy harvester based on self-oscillating switched-capacitor voltage doubler. IEEE J Solid-State Circuits. https://doi.org/10.1109/JSSC.2014.2346788

53. Jiang Y, Law M, Mak P, Martins RP (2018) Algorithmic voltage-feed-in topology for fully integrated fine-grained rational buck-boost switched-capacitor DC-DC converters. IEEE J Solid-State Circuits. https://doi.org/10.1109/JSSC.2018.2866929

Fully Integrated SC DC-DC Performance Enhancement Techniques State of the Art

4

4.1 Efficiency Enhancement Techniques

As discussed in Chap. 2 the efficiency of a Switched Capacitor (SC) converter strongly depends on the quality of its passive devices, namely on the flying capacitance parasitic value. Hence, techniques to decrease the parasitic capacitances of these, or to re-use the charge lost to them, have been investigated to increase the converter's efficiency. This section describes in detail the most commonly used techniques.

4.1.1 MOS Capacitor Parasitic Reduction Techniques

As discussed in Chap. 2, MOS capacitors offer the highest capacitance density in Complementary Metal-Oxide-Semiconductor (CMOS) bulk technology. However, due to their proximity to the substrate, they are also the ones with the highest parasitic values. Hence, techniques have been developed to decrease the influence of the parasitic capacitances [1–4]. Figure 4.1a shows the MOS capacitor with the drain/source shorted to the bulk [1]. This eliminates the drain/source to N-well junction capacitance, also named channel capacitance, C_c, and thus the parasitic capacitance is equal to the capacitance between the N-well and P-sub, which yields a lower parasitic value, resulting in $\alpha = 1.26\%$. In [2], instead of shorting the channel capacitance, C_c, a resistor is placed in series and the well is biased with the highest available voltage, in this case V_{IN}. As shown in Fig. 4.1b, if R_{BIAS} is high enough, the parasitic capacitance is now given by the series of the channel capacitance and the N-well to P-sub capacitance, resulting in $\alpha \approx 1.2\%$. Figure 4.1c shows the technique developed in [3], where a PMOS capacitor implemented with a deep N-well is presented. This reduces the channel capacitance, due to the low impedance between the P$^+$ and the P-well, and reduces the well-junction between the P-well and the N-well, since it is proportional to the N-well

© The Author(s), under exclusive license to Springer Nature Switzerland AG 2022
R. Madeira et al., *Fully Integrated Switched-Capacitor PMU for IoT Nodes*, Synthesis Lectures on Engineering, Science, and Technology,
https://doi.org/10.1007/978-3-031-14701-2_4

biasing voltage. Hence, a voltage doubler is used to generate $2V_{IN}$ for biasing the N-well. The previous two techniques were combined in [4] (Fig. 4.1d) and instead of a resistor, it uses to front-to-front diode-connected PMOS devices (Z_{BIAS}), to achieve high impedance with the minimum area overhead. In this the parasitic capacitances can be made smaller than 1% ($\alpha < 1\%$).

4.1.2   Charge Recycling

The top (αC_{FLY}) and bottom (βC_{FLY}) parasitic capacitances are one of the main reasons for the decrease in the converter's efficiency. Figure 4.2 shows the schematic of two Series-Parallel (SP) SC converters in phase opposition, a topology that has already been described and analysed in Chap. 2, without considering the charge recycling technique.

Figure 4.2a and c show converter configuration in ϕ_1 and ϕ_2, respectively, where during ϕ_1, C_{FLY} charges to $V_{IN} - V_{OUT}$, and during ϕ_2, C_{FLY} discharges to V_{OUT}. Hence, the top (bottom) parasitic capacitance, during ϕ_1, is charged to V_{IN} (V_{OUT}), and during ϕ_2, it is discharged to V_{OUT} (ground). Whilst the charge in the top parasitic capacitance is transferred to the output, the charge in the bottom parasitic capacitance is wasted. However, since there are two converters running in phase opposition, their parasitic capacitances can be averaged out, so that the effort of charging the parasitic capacitance is reduced [5]. This averaging takes place in the dead-time between phases ϕ_3, as shown in Fig. 4.2b, which is short to minimize the impact in C_{FLY}. This is not a problem, since the parasitic capacitances are much smaller than C_{FLY}, and thus, their time constant is much smaller. This technique can improve performance up to $\approx 2.5\%$ [5].

4.1.2.1 Scalable Parasitic Charge Redistribution

The previous technique only allows for a parasitic charge reduction of two, since only the voltage of two parasitic capacitors is averaged. Hence, if this averaging is to be increased to the point of almost reaching the voltage of the next phase, then the charge lost from the voltage supply would be minimum. This is the main idea behind the Scalable Parasitic Charge Redistribution (SPCR) technique [6]. It uses these two cells in phase opposition, however now divided in N interleaved cells, where the parasitic' charges are redistributed between them, in additional intermediary steps between the phase transitions.

Figure 4.3 shows the behaviour of an 1/2 SP SC with 8 interleaved cells and with 3 Charge Redistribution Step (CRS), for the re-use of the charge of the bottom plate parasitic capacitance, hereby name BP, for simplicity. The diagram and plot show the voltage at the bottom plate, that is equal to V_{OUT} on ϕ_1 and equal to V_{SS} on ϕ_2. Hence between this transition, each BP charging is paired with the BP with the closest higher voltage. The voltage will average out bringing the BP voltage closer to the highest voltage V_{OUT}. In fact, in the last transition to the highest voltage, the BP will be only $1/(CRS + 1)$ times the highest voltage

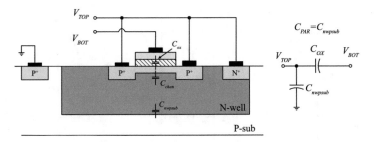

(a) Inversion-based capacitor with bulk connection shorted to $V_{Drain/Source}$ [1].

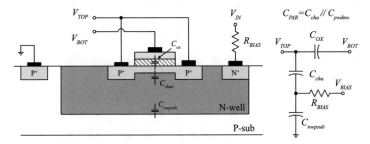

(b) Inversion-based capacitor with V_{BIAS} through R_{BIAS} [2].

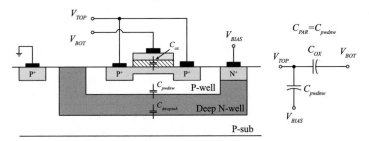

(c) Accumulation-based capacitor with V_{BIAS} [3].

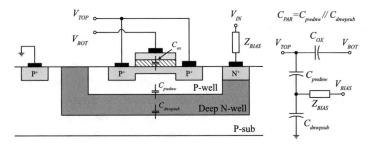

(d) Accumulation-based capacitor with V_{BIAS} through Z_{BIAS} [4].

Fig. 4.1 PMOS capacitor layout and equivalent schematic with biasing techniques [1–4]

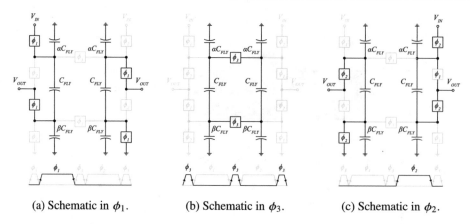

(a) Schematic in ϕ_1.   (b) Schematic in ϕ_3.   (c) Schematic in ϕ_2.

Fig. 4.2 Simplified schematic of a SP 1/2 SC converter with charge recycling

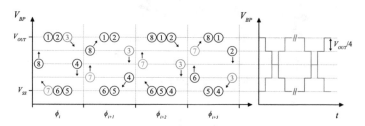

Fig. 4.3 Diagram of the charge redistribution process [6]

away from the highest voltage. This means the charge required to charge the BP to V_{OUT} is 4 times smaller, leading to 4 times less power dissipation. Furthermore, because the cells change phase sequentially, the order required for the connections for the BP capacitor is known at the design time.

4.1.3   Stage Outphasing

The previous techniques are able to improve the efficiency at low power densities. This is because the charge taken by the parasitic capacitances does not depend on the converter's absolute output power. The next techniques aim to improve the effective capacitance density by reducing the charge-sharing losses of the flying capacitors, hence also having an impact at high power densities [7]. The main objective of Stage Outphasing (SO) technique is to reduce the voltage variation between the top plates (TP) of the flying capacitors and thus reducing the power sharing losses. In the conventional Dickson converter showed in Chap. 3 Fig. 3.4, the converter cells are fully charged after the charging state and fully discharged after the discharging state. Meaning that a certain amount of charge, q, is transferred from the node

Fig. 4.4 SO applied to a
Dickson converter [4]

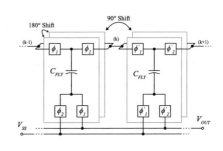

Fig. 4.5 Charge transfer
between two adjacent stages of
a Dickson converter using
SO [4]

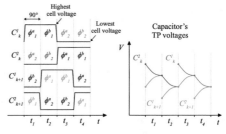

$k - 1$ to k. This charger transfer generates a voltage swing, ΔV, which is proportional to q, resulting in charge-sharing losses, P_{CS}, given by

$$P_{CS} \propto \frac{q^2}{C_k} \tag{4.1}$$

Figure 4.4, shows the application of the SO technique. Each stage is divided into two cells connected to the same nodes, $(k - 1)$ and (k), that run with a 180° phase shift. This means that there is always one, and only one, cell connected to one each node $(k - 1)$ and (k). By adding a phase shift of 90° to the next stage, then, one cell is going to connect two cells of the next (and previous) stage, during the charging or discharging state.

To understand the concept, Fig. 4.5, shows two cells, C_k^1 and C_{k+1}^1, and its respective opposite phase cells, C_k^2 and C_{k+1}^2, with their clock phases ϕ_1 and ϕ_2, further divided into two sub-phases (e.g. ϕ_1^a, ϕ_1^b). The capacitance values are equal between cells. Note that, C_k only connects to C_{k+1} when it is on ϕ_2 and C_{k+1} only connects to C_k on ϕ_1. For example, C_k^1 during ϕ_2 connects to C_{k+1}^1 and to C_{k+1}^2. Hence, at the beginning of t_3 which corresponds to the maximum TP voltage of C_k^1, the C_k^1 TP voltage will be averaged with the C_{k+1}^1 TP voltage. A the end of t_4, the TP voltage of C_k^1 is averaged out with C_{k+1}^2, reaching its minimum voltage. This behaviour is the same for all cells, thus, during ϕ_1, $V_{\phi_1^a} < V_{\phi_1^b}$, and the opposite during ϕ_2 ($V_{\phi_2^a} > V_{\phi_2^b}$). Hence, the cells of ϕ_2^b connect to cells of ϕ_1^a. The same goes for ϕ_2^a, which connects to ϕ_1^b, which have the highest voltage. This means that each cell always connects to the cell with the closest voltage.

The amount of charge transferred during the sub-phases is the same. Therefore, the amount of charge transferred during one phase is doubled. This is the equivalent of doubling the total flying capacitance, reducing the power sharing losses by a factor of 2 (4.2).

$$P_{CS} \propto 2\,C_k \left(\frac{\Delta V}{2}\right)^2 = \frac{q^2}{2\,C_k} \tag{4.2}$$

This technique is however more sensitive to parasitic capacitances of the flying capacitors due to the voltage swing in the intermediary nodes, which is DC at the conventional Dickson topology. Nonetheless, this impact may not be critical since the voltage swings in these nodes are small.

4.1.4   Multiphase Soft-Charging

It is easy to understand that if the number of cells per stage is increased, the number of charge transfers is increased as well. Therefore, each cell is divided into M cells operating with a clock phase that is phase-shifted $360°/M$ and the adjacent cells are further shifted $180°/M$ (Fig. 4.6). This creates M separated intermediate—$(k, 1)$ to (k, M). A full cycle, charge and discharge state, takes a total of $2M$ phases, which are split evenly in both states.

Figure 4.7 shows a simplified diagram of the charge transfers using the Multiphase Soft-Charging (MSC) technique. Following the same analysis as in SO, it can be shown that the charge transferred between two adjacent phases has the same value. Therefore, the charge transfer of all phases in a state must also be equal. Resulting in equidistant voltage values of the intermediate notes, $(k, 1)$ to (k, M). Hence, the power sharing losses are now divided by a factor of M, as shown in the following equation.

$$P_{CS} \propto \frac{q^2}{M\,C_k} \tag{4.3}$$

Fig. 4.6 MSC applied to a Dickson converter [4]

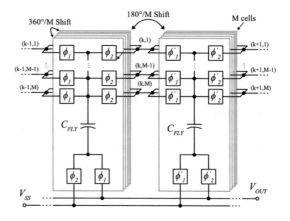

Fig. 4.7 Charge transfer
between two adjacent stages of
a Dickson converter using
MSC [4]

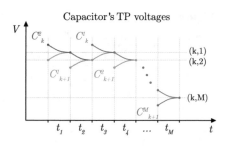

4.1.5   Full Soft-Charging Converter

The two above techniques, SO and MSC, can be combined by doubling the number of cells
per stage of the MSC and halve the phase shift between stages (Fig. 4.8). And thus, further
reducing the power losses by a factor of 2, as shown in the equation below.

$$P_{CS} \propto \frac{q^2}{2\,M\,C_k} \tag{4.4}$$

Both of these two techniques can only be applied to the charge transfer between capacitors.
In the case of a single capacitor that charges to a DC voltage, these techniques cannot be
applied.

4.1.6   Conclusions

Table 4.1 shows a table of the SC converters using efficiency enhancement techniques presented above. It is clear that these techniques have been pushing the converter's efficiency
closer to 100%. As for modifications in the biasing of the MOS capacitor, the resulting
efficiency values are not far from those presented in the previous chapter. The main increase
in the efficiency is found in the charge recycling techniques [5, 6]. Where the one presented
in [6] achieves the best result. However, the C_{FLY} is implemented with Metal-Oxide-Metal

Fig. 4.8 Combination of SO
and MSC applied to a Dickson
converter with a factor of M [4]

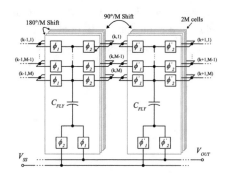

Table 4.1 Comparison table of SC converters using efficiency enhancement techniques

Publication	[1]'11	[2]'13	[3]'15	[4]'17	[5]'15	[6]'17
Technique	Flying well	Inversed-based C_{FLY} with R_{BIAS}	Accumulation-based C_{FLY}	Accumulation-based C_{FLY} with R_{BIAS}	Charge-Recycle	FSC
Tech. (nm)	90	65	65	28	31 SOI	40
CR #	1	2	2	1	1	1
CR	Down: 1/2	Down: 1/3, 2/5	Down: 1/4, 1/3	Down: 1/3	Down: 1/2	Down: 1/2
Switch #	4	15	8	4	8	24
C_{FLY} (Type)	MOS	MOS	MOS, MOM MIM	MOM+MOS	Deep-Trench	MOM
C_{OUT} (Type)	N.R	None	N.R.	None	SMD	None
C_{OUT} (nF)	N.R	–	N.R.	0	33	0
Phase interleaved	Yes (21)	Yes (18)	Yes (9)	Yes (2)	Yes (2)	Yes (16)
V_{IN} (V)	2.35–2.6	3–4	1.5–2.5	3.2	1.8	1.855–2.07
V_{IN} range (V)	2.4	1	1	–	–	0.22
V_{OUT} (V)	1	1	0.4–0.7	0.95	0.84	0.9
V_{OUT} range (V)	–	–	0.3	–	–	–
V_{Ripple} (mV)	80	297	N.R.		N.R.	18
Area (mm^2)	2.14	0.64	0.23	0.117	0.003	2.4
η_{max} (%)	69	74.3	79.5	82	86	94
P_{OUT} at η_{max} (mW)	900	122	26	12.9	15.6	3.15
$P_{density}$ at η_{max} (mW/mm^2)	421^2	40*	33.6	1.1	4.6^1	1.3
$P_{density_{max}}$ (mW/mm^2)	770^2	190	170*	1.1	4.6^1	1.3

*Estimated from the corresponding literature
[1]Estimation without the C_{OUT} area
[2]External regulation. N.R not reported

(MOM) capacitors, which have lower parasitic capacitances. Hence, the efficiency with MOS capacitors would be lower. Moreover, these charge redistributions were only applied to single Conversion Ratio (CR) converters. Where the main reason may be the circuit complexity that multi-ratio topologies would require in the charge redistribution matrix schemes.

4.2   Ripple Reduction Techniques

As explained in Sect. 2.4 of Chap. 2, the voltage ripple is caused either by not supplying charge fast enough to the output, and thus the output decoupling capacitor discharges below the target voltage, or by transferring more charge to the output than required. Whilst the first is solved by using a single bound hysteretic control with a fast response, the second can be minimized by using time interleaving cells. However, using the hysteretic control, the converter always work at the maximum clock frequency, and the comparator controls if the clock is fed to the converter or not, according to the output voltage. Hence, the amount of charge transferred is always the same. At lower frequency values, this will result in higher voltage ripples because the load impedance is much higher than the one at maximum power. In this sense, fine tuning mechanisms have been developed to minimize the ripple when working below the maximum frequency.

One way of controlling the voltage ripple is using pulse modulation [8, 9]. Lets consider the 1/1 SP SC converter of Fig. 4.9a. In ϕ_1, C_{FLY} is charged to V_{IN}. Hence, by decreasing the phase duration, the amount of charge transferred to C_{FLY} can be reduced. Figure 4.9b shows the output and phases waveforms for two different ϕ_1 values. For the same switch ON resistance and C_{FLY} value, if the charging time is smaller, then the charge transfer to the output will also be smaller, resulting in a smaller peak voltage at the output, hence minimizing V_{Ripple}.

The switch modulation [10–12] consists on controlling the switch's ON resistance value, and thus controlling the amount of charge transferred. Notice that this is only true if the clock phases duration is close to the circuit's RC time constant. This can be achieved in two ways—controlling the switches' gate-to-source voltage, V_{GS}, or by add/remove switches in parallel. Figure 4.10 shows the switch modulation technique applied through the control of V_G given that R_{ON} is inversely proportional it. Another way to control the switches' R_{ON} is to physically place N transistors in parallel, where the number of ON switches will vary the total transistor's width, and thus varying the R_{ON} value, as shown in Fig. 4.11.

Capacitance modulation [4, 6, 8, 13–15] is the most used charge regulation method as a secondary loop regulation, for ripple reduction. This controls the amount of charge transferred to the output by modifying the C_{FLY} value. This can be done by placing N capacitors in parallel, however, it would require 2 extra switches in series to connect and

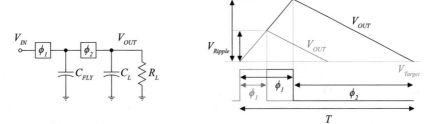

(a) Simplified schematic of the 1/1 SC converter.   (b) Voltage waveforms.

Fig. 4.9 Example of the pulse modulation technique [8]

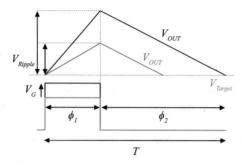

Fig. 4.10 Example of the switch modulation technique by controlling the switches' V_G [8]

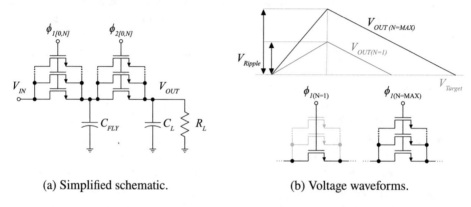

(a) Simplified schematic.   (b) Voltage waveforms.

Fig. 4.11 Example of the switch modulation technique by controlling the number of switches in parallel [10–12]

disconnect the capacitors. Hence, typically the converter is divided into N smaller converters, named cells, that are enabled or not according to the charge required at the output, as shown in Fig. 4.12. The gate driver's complexity remains almost the same since turning ON or OFF the cells can be done by either pass or not the phase signals to the cell. Nonetheless, since C_{FLY} is large, only a coarse tuning can be achieved, without using a large number of cells, and thus increasing the layout complexity of the cells [8].

4.2.1 Conclusions

The above techniques allow to minimize the output ripple voltage when the converter is not operating at the maximum clock frequency, i.e. is not operating at the maximum output power level. These techniques are typically used as a secondary control loop. The main goal is to ideally operate the converter always close to the maximum frequency and keep the charge constant [13]. The switch modulation typically increases the gate driver complexity

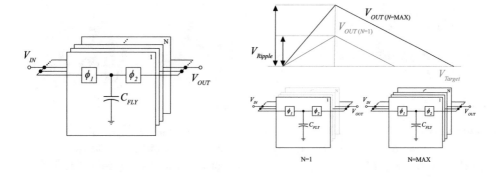

(a) Simplified schematic.   (b) Voltage waveforms.

Fig. 4.12 Example of the capacitance modulation technique by controlling the number of converter cells in parallel [4, 6, 8, 13–15]

and area, especially in the parallel switches techniques. As for capacitance modulation, if time interleaved is employed, the cells are already fragmented. Hence, they can easily be disabled by feeding or not the clock phases. Hence, all the time interleaved converters can easily be modified to apply capacitance modulation.

4.3 Dynamic Power Allocation Technique

In most multi-core application processors designs, SC converters with different power and voltage specifications are independently designed (Fig. 4.13). This requires a large area overhead, especially if both converters do not work simultaneously at full power. In recent years, multi-output multi-ratio SC converters have been designed to minimize the on-chip area overhead. In [16], a dual-output SC regulator is shown with two output voltages, one with a CR of 2/1 and another with 3/1. Even though both converters are combined, the two output voltages have fixed CRs and external flying and decoupling capacitors. In [17], two multi-ratio SC converters are presented, where one has an off-chip flying capacitor and the other on-chip capacitors, both with external decoupling. The converter with the off-chip capacitor is used to handle higher loads, i.e. low R_L values ($I_L = 10$ mA), and the on-chip lighter loads, i.e. high R_L values ($I_L = 1$ mA), where the output is chosen according to the power demand. This work has now a power on-demand control that enables to select of which

Fig. 4.13 Conventional Separated Converters with two voltage domains

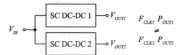

Fig. 4.14 Dual-Output SC Converter with dynamic power allocation, for $P_{OUT_1} > P_{OUT_2}$ (left) and $P_{OUT_1} < P_{OUT_2}$ (right)

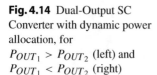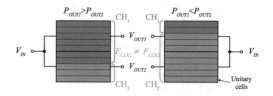

converter is used for which output node. However, both converters are separated and still use off-chip capacitors. In [18], three different SC converters are integrated – a multi-ratio 7-bit binary-reconfigurable SC ($V_{OUT} = 1.2$ V), a tippler Dickson step-up converter ($V_{OUT} = 3.3$ V), and a 1/2 SP step-down ($V_{OUT} = 0.6$ V). The binary-reconfigurable converter handles the input voltage variations and its output voltage is the input of both the tippler and the 1/2 step-down. The regulation is achieved using a lower-bound hysteretic controller for each converter, hence they run independently. Even though the controller is integrated, the three converters are separated and do not share resources. Moreover, the authors do not mention the use of time interleaved techniques, hence it is expected to have external output decoupling capacitors for each output. Despite the external capacitors, everything else is integrated. In the aforementioned works, there is no capacitance reallocation, hence the capacitor utilization is low. This is because the converters were designed to work at full power, hence, when working at lower power levels, the flying capacitance is oversized.

To achieve higher power density, the converters must be combined into one converter with multiple outputs. To this end, a technique named PDA (PDA) was developed in [19]. It consists on dividing a converter into 2 groups of N cells, which work in an interleaved way, and the number of cells for each output is allocated according to the power demand. Figure 4.14 shows the dynamic power allocation between two SC converters, for two different scenarios of output power. The two converters are divided into 2 channels (CH_1 and CH_2), that output two voltages. Each output is regulated through frequency modulation and, with the PDA technique, it is expected to be approximately the same for both the converters using cell modulation, i.e. by adding or subtracting cells to the channels. By making the clock frequency of both channels the same, balancing the power density and maximizing the capacitance utilization, both switching and parasitic losses are reduced, without compromising the power delivery to the loads. Hence, this maximizes the capacitance utilization leading to better power density and overall efficiency.

4.3.1 Conclusions

Table 4.2 shows the state-of-the-art comparison of the multi-output multi-ratio capacitors. The work [19] achieves the highest power density, making it clear the advantage maximizing capacitance utilization by using PDA.

Table 4.2 Comparison table of multi-output multi-ratio SC converters

Publication	[16]'15	[17]'16	[18]'16	[19]'17
Tech. (nm)	550	65	180	28
CR #	2	6	3[1]	2
CR	Up: 2/1, 3/1	Down: 1/3, 2/5, 1/2, 2/3, 1/1 Up: 3/2	Down: 1/2, 1/3, Binary Up: Binary	Down: 2/3, 1/2
Output #	2	2	3	2
C_{FLY} (Type)	SMD	MOS, SMD	MOS, MIM	MOM, MOS
C_{OUT} (Type)	SMD	SMD	N.R.	N.R.
C_{OUT} (nF)	N.R	N.R.	N.R.	N.R.
V_{IN} (V)	1.1–1.8	0.85–3.6	0.9–4	1.3–1.6
V_{IN} range (V)	0.7	2.75	3.1	0.3
V_{OUT} (V)	2, 3	0.1–1.9	0.6, 1.2, 3.3	0.4–0.9
V_{OUT} range (V)	–	0.8	–	0.5
V_{Ripple} (mV)	56	N.R.	N.R.	N.R.
Area (mm^2)	6.9	6	2.36	0.6
η_{max} (%)	89.5	95.8	81	83.3
P_{OUT} at η_{max} (mW)	2.5	12	0.2	45
$P_{density}$ at η_{max} (mW/mm^2)	N.A.	2[2]	0.08*	75
$P_{density_{max}}$ (mW/mm^2)	N.A.	2[2]	0.25	150

* Estimated from the corresponding literature
[1] Separated converters - 7b Binary + 1/2 SP + 1/3 Dickson
[2] Estimated considering only the power for the integrated converter. N.R. Not reported, N.A. Not applicable

4.4 Distributive Clock Generation

In traditional designs, the clock phases for the interleaved cells are generated on the clock generator, and then distributed across the chip to the corresponding cells. Even though the line delay is not critical, when using a large number of phases in a high frequency converter, it can become a problem. To minimize this, the phases can be generated locally, on each converter cell, by a delay circuit [20–22]. Hence, each cell passes the clock to the next cell. This reduces the complexity of the clock generator and allows for a more modular structure for the cells. Figure 4.15 shows the conceptual layout of the multi-phase converter ring proposed in [20–22]. In this, the converter cells are laid out around the load circuits in

Fig. 4.15 Ring-shaped multi-phase SC converter with on-chip V_{OUT} and V_{SS} grids [21, 22]

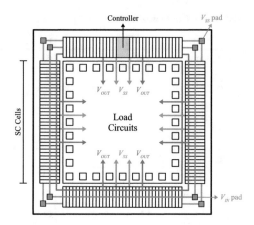

a ring shape. Each unitary cell is composed of a three-stage SP SC converter that can be configured in the 1/2, 2/3, and 3/4 CR. Each cell passes the clock signal to the next cell in an interleaved manner, giving a total of 123 interleaved cells. This work has reported a high maximum power density value of 180 mW/mm^2 and a maximum efficiency of 78.3%. The downside of this technique is that the delay on one cell causes a delay in all the cells, hence, it could cause a voltage droop on the output voltage if the delay is significant. Furthermore, at high frequencies, the parasitics of the lines must be taken into accounted in the circuit delay design. Also, Process, Voltage, and Temperature (PVT) variations and mismatch effects are increased since the delay cells are distributed along the chip. Nonetheless, these distributive clock schemes bring layout flexibility and can decrease the power consumption of the clock generation, when compared to centralized clock generation schemes [23].

References

1. Meyvaert H, Breussegem TV, Steyaert M (2011) A monolithic 0.77W/mm^2 power dense capacitive DC-DC step-down converter in 90nm Bulk CMOS. In: European solid-state circuits conference (ESSCIRC). https://doi.org/10.1109/ESSCIRC.2011.6045012
2. Le H, Crossley J, Sanders SR, Alon E (2013) A sub-ns response fully integrated battery-connected switched-capacitor voltage regulator delivering 0.19W/mm^2 at 73% efficiency. In: IEEE international solid-state circuits conference (ISSCC). https://doi.org/10.1109/ISSCC.2013.6487775
3. Jiang J, Lu Y, Huang C, Ki W, Mok PKT (2015) A 2-/3-phase fully integrated switched-capacitor DC-DC converter in bulk CMOS for energy-efficient digital circuits with 14% efficiency improvement. In: 2015 IEEE international solid-state circuits conference (ISSCC). https://doi.org/10.1109/ISSCC.2015.7063078
4. Butzen N, Steyaert MSJ (2017) Design of soft-charging switched-capacitor DC-DC converters using stage outphasing and multiphase soft-charging. IEEE J Solid-State Circuits. https://doi.org/10.1109/JSSC.2017.2733539

5. Andersen TM, Krismer F, Kolar JW, Toifl T, Menolfi C, Kull L, Morf T, Kossel M, Brändli M, Buchmann P, Francese PA (2013) A 4.6 W/mm^2 power density 86% efficiency on-chip switched capacitor DC-DC converter in 32 nm SOI CMOS. In: 2013 twenty-eighth annual ieee applied power electronics conference and exposition (APEC). https://doi.org/10.1109/APEC.2013.6520285

6. Butzen N, Steyaert M (2016) Scalable parasitic charge redistribution: design of high-efficiency fully integrated switched-capacitor DC-DC converters. IEEE J Solid-State Circuits. https://doi.org/10.1109/JSSC.2016.2608349

7. Butzen N, Steyaert M (2019) Advanced Multiphasing: Pushing the Envelope of Fully Integrated Power Conversion. IEEE Custom Integrated Circuits Conference (CICC). https://doi.org/10.1109/CICC.2019.8780134

8. Kudva SS, Harjani R (2013) Fully integrated capacitive DC-DC converter with all-digital ripple mitigation technique. IEEE J Solid-State Circuits. https://doi.org/10.1109/JSSC.2013.2259044

9. Lee H, Hua Z, Zhang X (2015) A reconfigurable $2 \times /2.5 \times /3 \times /4\times$ SC DC–DC regulator with fixed on-time control for transcutaneous power transmission. In: IEEE transactions on very large scale integration (VLSI) systems. https://doi.org/10.1109/TVLSI.2014.2315158

10. Kwong J, Ramadass Y, Verma N, Koesler M, Huber K, Moormann H, Chandrakasan A (2008) A 65nm Sub-Vt microcontroller with integrated SRAM and switched-capacitor DC-DC converter. In: IEEE international solid-state circuits conference (ISSCC). https://doi.org/10.1109/ISSCC.2008.4523185

11. Kwong J, Ramadass YK, Verma N, Chandrakasan AP (2009) A 65 nm Sub-V_t microcontroller with integrated SRAM and switched capacitor DC-DC converter. IEEE J Solid-State Circuits. https://doi.org/10.1109/JSSC.2008.2007160

12. ianxi L, Hao C, Tianyuan H, Junchao M, Zhangming Z, Yintang Y (2018) A dual mode step-down switched-capacitor DC-DC converter with adaptive switch width modulation. Microelectron J. https://doi.org/10.1016/j.mejo.2018.06.003

13. Breussegem TV, Steyaert M (2010) A fully integrated gearbox capacitive DC/DC-converter in 90nm CMOS: optimization, control and measurements. In: IEEE 12th workshop on control and modeling for power electronics (COMPEL). https://doi.org/10.1109/COMPEL.2010.5562379

14. Ramadass YK, Fayed AA, Chandrakasan AP (2010) A fully-integrated switched-capacitor step-down DC-DC converter with digital capacitance modulation in 45 nm CMOS. IEEE J Solid-State Circuits. https://doi.org/10.1109/JSSC.2010.2076550

15. Bang S, Seo J, Chang L, Blaauw D, Sylvester D (2016) A low ripple switched-capacitor voltage regulator using flying capacitance dithering. IEEE J Solid-State Circuits. https://doi.org/10.1109/JSSC.2015.2507361

16. Hua Z, Lee H (2015) A reconfigurable dual-output switched-capacitor DC-DC regulator with sub-harmonic adaptive-on-time control for low-power applications. IEEE J Solid-State Circuits. https://doi.org/10.1109/JSSC.2014.2379616

17. Teh CK, Suzuki A (2016) A 2-output step-up/step-down switched-capacitor DC-DC converter with 95.8% peak efficiency and 0.85-to-3.6V input voltage range. In: IEEE international solid-state circuits conference (ISSCC). https://doi.org/10.1109/ISSCC.2016.7417987

18. Jung W, Gu J, Myers PD, Shim M, Jeong S, Yang K, Choi M, Foo Z, Bang S, Oh S, Sylvester D, Blaauw D (2016) A 60%-efficiency 20nW-500μW tri-output fully integrated power management unit with environmental adaptation and load-proportional biasing for IoT systems. In: IEEE international solid-state circuits conference (ISSCC). https://doi.org/10.1109/ISSCC.2016.7417953

19. Jiang J, Lu Y, Ki W, U S, Martins RP (2017) A dual-symmetrical-output switched-capacitor converter with dynamic power cells and minimized cross regulation for application processors in 28nm CMOS. In: IEEE international solid-state circuits conference (ISSCC). https://doi.org/10.1109/ISSCC.2017.7870402

20. Lu Y, Jiang J, Wing-Hung K, Yue CP, Sai-Weng S, Seng-Pan U, Martins RP (2015) A 123-phase DC-DC converter-ring with fast-DVS for microprocessors. In: IEEE international solid-state circuits conference (ISSCC). https://doi.org/10.1109/ISSCC.2015.7063077

21. Lu Y, Jiang J, Ki W (2018) Design considerations of distributed and centralized switched-capacitor converters for power supply on-chip. IEEE J Emerg Select Topics Power Electron. https://doi.org/10.1109/JESTPE.2017.2747094

22. Lu Y, Jiang J, Ki W (2017) A multiphase switched-capacitor DC-DC converter ring with fast transient response and small ripple. IEEE J Solid-State Circuits. https://doi.org/10.1109/JSSC.2016.2617315

23. Jiang J, Liu X, Ki WH, Mok PKT, Lu Y (2021) Circuit techniques for high efficiency fully-integrated switched-capacitor converters. IEEE Trans Circuits Syst II: Express Brief. https://doi.org/10.1109/TCSII.2020.3046514

Design of a Fully Integrated Power Management Unit

5

5.1   Overview

This chapter describes the design process of a 16 mW fully integrated Power Management Unit (PMU), implemented in a bulk 130 nm Complementary Metal-Oxide Semiconductor (CMOS) technology. Figure 5.1 shows the simplified block diagram of the proposed PMU, where a variable voltage of a supercapacitor is converted into a stable voltage of 0.9 V, suitable for power systems like Internet of Things (IoT) nodes. The PMU includes a 1 + 3 binary-weighted multi-ratio multi-cell Switched Capacitor (SC) converter, and a set of auxiliary circuits required to ensure the correct behaviour of the converter, according to the supercapacitor's voltage and the output load. These auxiliary circuits are composed of the phase generator, Conversion Ratio (CR) controller, cell controller, switch's drivers, reference voltage generator, and the start-up circuits. These are powered by the converter and can supply information to IoT system about the PMU's operation and the energy available in the supercapacitor. Each of these blocks is also described in this chapter. The main goal was to integrate all these blocks while maximizing the PMU's efficiency and the energy extracted from the supercapacitor.

5.2   Multi-ratio Multi-cell Fully Integrated SC DC-DC Converter Design

5.2.1   Topology and Conversion Ratio Selection

As described in Chap. 3, the Series-Parallel (SP) topology can achieve a good compromise between high power density and high efficiency. Hence, it was the chosen topology to implement the multi-ratio SC converter. As shown in [1], with one flying capacitor, three

© The Author(s), under exclusive license to Springer Nature Switzerland AG 2022
R. Madeira et al., *Fully Integrated Switched-Capacitor PMU for IoT Nodes*, Synthesis Lectures on Engineering, Science, and Technology,
https://doi.org/10.1007/978-3-031-14701-2_5

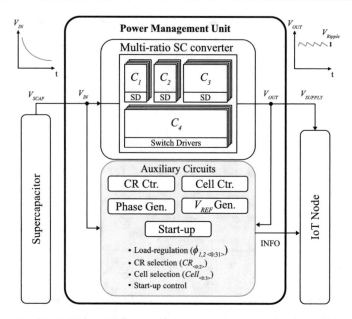

Fig. 5.1 Simplified block diagram of the PMU used to supply power to an IoT system through the energy of a supercapacitor

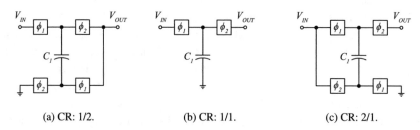

(a) CR: 1/2.   (b) CR: 1/1.   (c) CR: 2/1.

Fig. 5.2 SP converters that can be implemented with one flying capacitor [1]

CRs can be implemented, depicted in Fig. 5.2. The step-down 1/1, 1/2 and the step-up 2/1 CRs. Figure 5.3 shows the converter's efficiency as a function of V_{IN}, for $V_{OUT} = 0.9$ V and with the C_{FLY} implemented by a MOS capacitor. As expected the efficiency is only high near the V_{OUT}/CR value, hence in the transitions between CRs, the efficiency drops to values around 50% or below, thus decreasing the average efficiency value. Hence more CRs are required, meaning that the number of flying capacitors must be increased. With two capacitors, four extra CRs can be implemented, as shown in Fig. 5.4, in addition to the previous three CRs, by placing the two flying capacitors in parallel. The new CRs are the step-down 2/3 and 1/3, and the step-up 3/1 and 3/2. As shown in Fig. 5.5, for $V_{OUT} = 0.9$ V, the step-down 1/3 CR covers a voltage above the maximum 2.7 V of the supercapacitor, hence will not be used. With the 2/3 CR, the minimum efficiency stays above 65% for the 0.9–2.7 V input range. As for the step-up 3/2, 2/1 and 3/1 CRs, they produce a constant

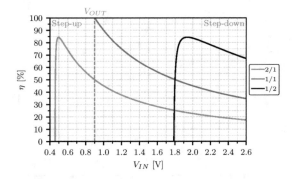

Fig. 5.3 SP converters efficiency as a function of V_{IN}, with one flying capacitor, for $V_{OUT} = 0.9$ V and with C_{FLY} implemented by a PMOS capacitor

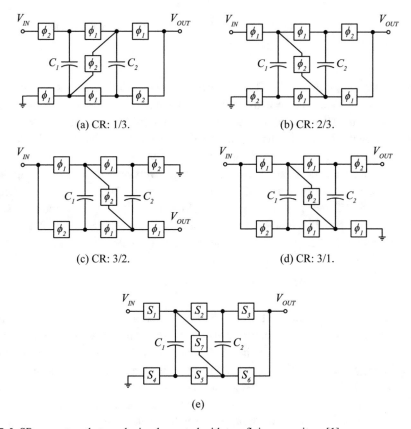

(a) CR: 1/3.

(b) CR: 2/3.

(c) CR: 3/2.

(d) CR: 3/1.

(e)

Fig. 5.4 SP converters that can be implemented with two flying capacitors [1]

output voltage for the input range of 0.9–0.3 V. However, there are some constraints that must be taken into account. First, the transition between the 1/1 and 3/2 CR will have a

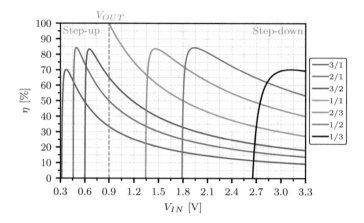

Fig. 5.5 SP converter's efficiency as a function of V_{IN}, with two flying capacitors, for $V_{OUT} = 0.9$ V and if C_{FLY} were implemented by an MOS capacitor

minimum efficiency below 60%. This is because the 1/1 CR will not work until 0.9 V, and the transition will have to occur in the 0.95–1.1 V input range, due to the high frequency of operation required. Hence there is a gap between both CRs. Second, the 3/2 CR, as shown in Fig. 5.4c, the flying capacitor has a voltage inversion, when compared with the step-down topologies. Furthermore, the input voltage is close to the output voltage, hence the voltage swing of the capacitor will be small. Both the voltage inversion and the small voltage swing of the capacitor prevent the use of MOS capacitors. Hence, if the 3/2 CR is not used, the gap between the 1/1 and 2/1 CR will bring the efficiency down to values around 50%.

Since the main goal is to efficiently extract the energy of a supercapacitor, it is important to see how much energy each topology is able to extract. The energy of a capacitor is given by

$$\Delta E_C = C \int_{V_i}^{V_f} V_c \, dV_c = \frac{1}{2} C \left(V_f^2 - V_i^2 \right) \tag{5.1}$$

Table 5.1 shows the input voltage range of operation and the extracted energy in the ideal case, i.e. if the converter efficiency was 100% throughout the whole V_{IN} range, and the extracted energy considering the mean efficiency of each CR. The numbers of the ideal extracted energy show that the gross of the energy is transferred in the step-down converters, approximately 88.9%, while the step-up converters only extract approximately 9.8%. Considering the converter's efficiency numbers, the extracted energy values decrease to 68.6% and 7.4%, respectively. This shows the discrepancy of the energy extracted by the step-down and the step-up converters when extracting the energy of a supercapacitor to produce an output voltage of 0.9 V. With this in mind, and knowing that implementing the 3/2 CR would prevent using MOS capacitors, lowering the power density per area, only the step-down 1/2, 2/3, and 1/1 CRs were selected to be merged into a multi-ratio SC converter.

The next step is to decide how to merge the three topologies. Two options were considered as shown in Fig. 5.6. Both have two flying capacitors C_1 and C_2 that are configured into

Table 5.1 Extracted energy of a supercapacitor numbers with SP converters using two flying capacitors, for a $V_{OUT} = 0.9$ V and with C_{FLY} implemented by a MOS capacitor

CR	V_{IN} (V)[a]	Ideal extracted energy (%)	Converter's efficiency (%)	Extracted energy (%)
1/2	1.822–2.7	54.46	75.68	41.22
2/3	1.364–1.822	20.02	78.68	15.75
1/1	0.9–1.364	14.41	80.46	11.59
3/2	0.614–0.9	5.94	75.44	4.48
2/1	0.462–0.614	2.24	79.14	1.78
3/1	0.3–0.462	1.69	68.51	1.16

[a]The voltage limits were chosen so that the efficiency is maximized, hence the converters change when the efficiency of the next, or previous, CR is higher

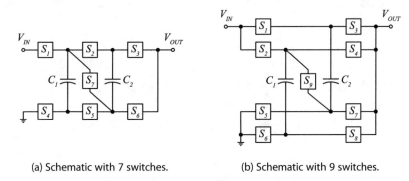

(a) Schematic with 7 switches.   (b) Schematic with 9 switches.

Fig. 5.6 Proposed SP multi-ratio SC converter simplified schematics

the three CRs by seven switches in Fig. 5.6a and 5.9 switches in Fig. 5.6b. The number of switches should be kept minimum to keep the power dissipation of the switches as low as possible. Moreover, the number of switches in series should also be minimized to decrease the transistors width, by having higher R_{ON} values, and thus also keep the switch power dissipation as low as possible. With this in mind, the chosen converter was the one with nine switches (Fig. 5.6b) because it has always two switches in each capacitor path, with the exception for the 2/3 CR, when C_1 and C_2 are placed in series. Figure 5.7 shows the proposed multi-ratio converter configuration in each CR for each phase. The theoretical analysis, presented in Chap. 2, to determine the input/output currents, the output voltage, the clock frequency, and the input/output powers that determine the efficiency was carried out for each CR, and is presented in Appendix A.

The flying capacitors were implemented with MOS capacitors, which are the ones with the highest capacitance density in bulk CMOS 130 nm technology, to maximize the converter's power density. As explained in Sect. 2.1.1 of Chap. 2, the MOS capacitor requires the channel to be isolated from the substrate so that the transistor's well remains reverse biased to decrease the capacitance parasitics [2, 3]. Hence, PMOS capacitor was chosen over the

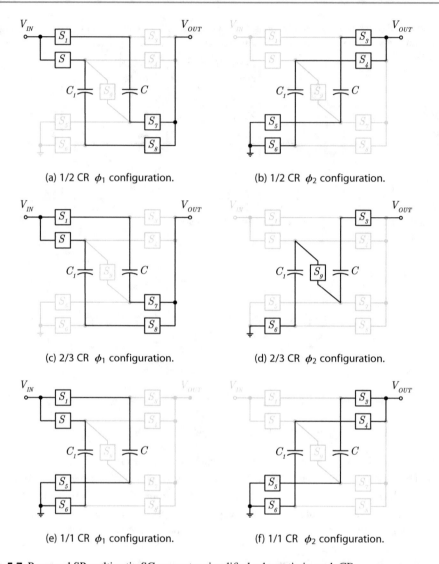

(a) 1/2 CR ϕ_1 configuration.   (b) 1/2 CR ϕ_2 configuration.

(c) 2/3 CR ϕ_1 configuration.   (d) 2/3 CR ϕ_2 configuration.

(e) 1/1 CR ϕ_1 configuration.   (f) 1/1 CR ϕ_2 configuration.

Fig. 5.7 Proposed SP multi-ratio SC converter simplified schematic in each CR

NMOS capacitor because it is already isolated from the substrate, not requiring the use of triple-well, which has a larger parasitic capacitance. Furthermore, the PMOS capacitor has a dominant top parasitic capacitance, shown in Sect. 2.3 of Chap. 2, that acts like an extra 1/1 converter in the 1/2 CR, thus increasing the capacitance making it possible to deliver more power for the same switching frequency. This has also a similar impact in the 2/3 and 1/1 CRs, as shown in Appendix A. In 130 nm CMOS technology, this value was empirically estimated to be higher than the one reported in [4], being approximately 3%.

5.2.2   Switch Selection

To determine the converter's efficiency including the switches' power dissipation effect, the transistors to implement each switch must be chosen. They must be selected having into account that almost all of them operate in the three CRs and without exceeding the maximum allowed voltage set by the technology. As seen in Chap. 2, 1.2 V transistors are preferable to 3.3 V transistors, and NMOS transistors are preferable to PMOS transistors. With that in mind, the switches that connect to V_{IN} ($S_{1,2}$), which is the highest voltage, were implemented with 1.2 V PMOS transistors, since NMOS transistors would require a high gate voltage to turn ON. To avoid exceeding the transistors' maximum allowed voltage the clock driver will be designed in a way to always generate V_{GS} within the maximum voltage limit. In 130 nm CMOS bulk technology, the oxide thickness of 1.2 V transistor is 2.7 nm. Hence, based on the oxide reliability projections from different research groups, shown in [8], the V_{GS} of the transistors was kept within 1.6 V, in order that no more than a specified failure rate will result. This is a value higher than the recommended V_{DD} plus 10%, however being this circuit a prototype, it was decided to use the maximum of 1.6 V to achieve the best performance possible. The switches that connect to ground ($S_{5,6}$) were implemented with 1.2 V NMOS transistors. The switches that connect to V_{OUT} were chosen to be implemented by 1.2 V PMOS transistors. Again, PMOS were chosen instead of NMOS because of the V_{GS} generation, which would increase the complexity of the clock driver. The switch S_9 only connects in one of the CRs and was chosen to be implemented with a 3.3 V NMOS transistor, because turning ON either a 1.2 V NMOS or a 1.2 V PMOS transistor, without exceeding their maximum voltage limits, would require using a low gate-to-source voltage, thus resulting in a very large transistor to achieve low R_{ON} value. Since this transistor only connects in the 2/3 CR, which will work in the input voltage range of 1.4–2 V, the input voltage can be used turn ON the 3.3 V NMOS. Concluding, the switches $S_{1,2,3,4,7,8}$ are implemented with 1.2 V PMOS transistors and the switches $S_{5,6}$ are implemented with a 1.2 V NMOS transistors, and lastly S_9 is implemented with 3.3 V NMOS transistor. The switches sizing and the resulting efficiency equations can be seen in Appendix A.

5.2.3   Efficiency and Conversion Ratio Limits

Figure 5.8 shows the converter's efficiency and F_{CLK} contour plot as a function of V_{IN} and the power density (P_{OUT}/Area), for $V_{OUT} = 0.9$ V and with C_{FLY} implemented by a PMOS capacitor (with a capacitance density of 4.76 fF/μm^2, $\alpha = 3\%$, and $\beta \approx 0\%$) and with the transistors sized using the 4τ methodology, whose equations can be seen in Appendix A. The main objective is to work as far right as possible in the power density axis and as far down as possible in the V_{IN}-axis. While keeping in mind the values of the efficiency and F_{CLK}. Hence, for the 1/2 CR, in the 1.9–1.85 V region, beyond the 10 mW/mm^2 value, the frequency starts to increase rapidly, and the efficiency starts to decrease. The same can be observed in the 2/3 CR for the region between 1.5 and 1.4 V. And finally the same for the

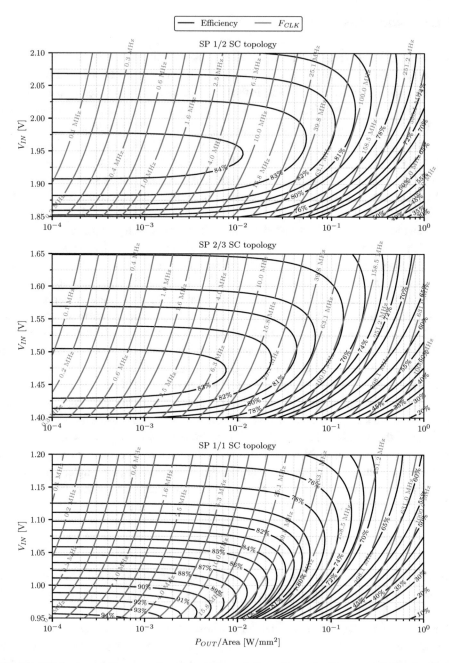

Fig. 5.8 Efficiency and F_{CLK} as a function of V_{IN} and P_{OUT}/Area, for $V_{OUT} = 0.9$ V and with the C_{FLY} implemented by an MOS capacitor

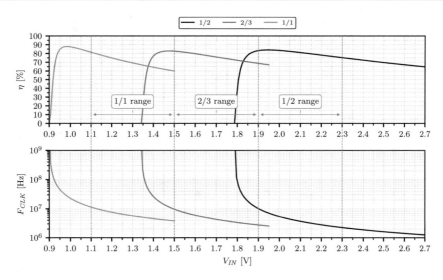

Fig. 5.9 Converter's efficiency and F_{CLK} as a function of V_{IN}, for P_{OUT}/Area= 10 mW, V_{OUT} = 0.9 V, and with C_{FLY} implemented by a PMOS capacitor

1/1 CR in the range of 1.10–0.9 V. Hence, this value was chosen for the converter design, meaning that the maximum F_{CLK} is around 10 MHz for all the three CRs.

Figure 5.9 shows the converter's efficiency and F_{CLK} graphs as a function of V_{IN}, for P_{OUT}/Area= 10 mW and in the previous conditions of Fig. 5.8. The graph shows that after 10 MHz the frequency starts to increase rapidly, which can result in a significant higher power dissipation of the clock circuitry for certain corners or even in a decrease of the output voltage. Therefore, the CR voltage thresholds are determined by the input voltage values where F_{CLK} becomes larger than 10 MHz. This allows to cover a V_{IN} range of 1.1– 2.3 V, with the voltage ranges for each topology of 1.1–1.5 V for the 1/1, 1.5–1.9 V for the 2/3, and 1.9–2.3 V for the 1/2 CR.

Table 5.2 shows the extracted energy of the supercapacitor considering the design previously described, with the supercapacitor energy given by (5.1), with an initial voltage of 0 V and a final voltage of 2.7 V. Given the converter's input operation voltage range of 1.1– 2.3 V, the ideal energy extracted from the supercapacitor is 55.98%. Considering the average efficiency of each converter's CR, the total energy extracted is approximately 42.67%. The previous extracted energy value from Table 5.1, for the step-down converters, was 68.6%. The difference between this value and final value (42.67%) comes mainly from the 2.3 V voltage limitation of the converter. Because the ideal extracted energy value reduces by a factor of 2.4, from 54.56% to 23.05%. The 2/3 and 1/1 CRs have an extracted energy value close to the one in Table 5.1. The remaining energy in the supercapacitor, below 1.1 V, is approximately 16.6%, which is higher than the 9.8% from Table 5.1. Nonetheless, the 1/1 converter cannot work until 0.9 V. If one considered the limit of 1.0 V then the ideal extracted energy of the supercapacitor would be 13.7%, which is not far from the current value, 16.6%.

Table 5.2 Extracted energy of a supercapacitor with the proposed multi-ratio SC converter, for a $V_{OUT} = 0.9$ V, P_{OUT}/Area= 10 mW/mm^2, and with C_{FLY} implemented by a PMOS capacitor

CR	V_{IN} (V)	Ideal extracted energy (%)	Converter's efficiency (%)	Extracted energy (%)
1/2	1.9–2.3	23.05	80.60	18.57
2/3	1.5–1.9	18.66	75.89	14.16
1/1	1.1–1.5	14.27	69.65	9.94

5.2.4 Time Interleaved and Capacitance Modulation

The output decoupling capacitor must be integrated so that the PMU is fully integrated. However, to keep the output voltage ripple low, its value must be hundred times higher than the flying capacitors. Thus, as explained in Chap. 2, time interleaved [3, 9–23] can be employed to reduce the output decoupling capacitor value suitable of being integrated. Furthermore, since the converter is going to be divided into N smaller cells, capacitance modulation [6, 11, 24–27] can be employed to reduce the capacitance value when working far from the maximum frequency point of operation, e.g. when the output power is lower than the maximum output power value. The output power can be sensed by adding one or more extra comparators to sense the output voltage [25, 26] or by sensing the average clock frequency value [6, 11, 24]. To avoid adding extra comparators, the sensing of the output power through the average clock frequency was chosen for the cells' activation and deactivation.

To explain how the cells can be sized, let's analyse again the 1/2 CR converter (Chap. 2, Sect. 2.3, Fig. 2.11a), where the clock frequency (F_{CLK}) equation is repeated were for convenience (5.2). This equation can be solved for C_{FLY}, resulting in the equation shown in (5.3).

$$F_{CLK} = \frac{P_{OUT}}{C_{FLY} V_{OUT} (V_{IN} (2 + \alpha) - V_{OUT} (4 + \alpha + \beta))} \tag{5.2}$$

$$C_{FLY} = \frac{P_{OUT}}{F_{CLK} V_{OUT} (V_{IN} (2 + \alpha) - V_{OUT} (4 + \alpha + \beta))} \tag{5.3}$$

To size the cells, the maximum capacitance value (C_{TOTAL}) required for working at the maximum frequency value (F_{max}), i.e. when P_{OUT} is at its maximum value (P_{max}), is determined by

$$C_{TOTAL} = \frac{P_{max}}{F_{max} V_{OUT} (V_{IN} (2 + \alpha) - V_{OUT} (4 + \alpha + \beta))} \tag{5.4}$$

The least significant cell is given by $C_{LSB} = C_{TOTAL}/(2^N - 1)$, where the minus one factor comes from the need of having one cell always active. Hence, the cells will be sized as $1\times, 2\times, 4\times, 8\times, \dots, 2^N \times C_{LSB}$, resulting in binary-weighted cells. The output

Fig. 5.10 Operation frequency range of the 1/2 CR SC converter with binary-weighted cells, for $C_{LSB} = 1.1$ nF, $P_{OUT_{max}} = 16$ mW, $F_{CLK_{max}} = 10$ MHz, $V_{IN} = 1.9$ V, $V_{OUT} = 0.9$ V, and including the parasitic capacitances from the PMOS capacitor for each cell

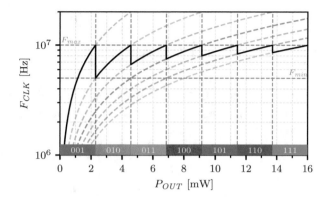

power becomes divided into $P_{LSB} = P_{max}/(2^N - 1)$. For example, considering $V_{OUT} = 0.9$ V, $V_{IN} = 1.9$ V, $P_{max} = 16$ mW, $F_{max} = 10$ MHz, $\alpha = 3\%$, and $\beta = 0\%$, then $C_{TOTAL} = 7.7$ nF, which using 3 bits, i.e. 3 cells, gives $C_{LSB} = 1.1$ nF and $P_{LSB} = 2.29$ mW. Figure 5.10 shows the plot of the F_{CLK} Eq. (5.2) for the binary-weighted cells (dashed lines), where each cell $C_{FLY<0:2>}$ has a capacitance value that is a binary-weighted multiple of $C_{LSB} = 1.1$ nF, for $P_{OUT_{max}} = 16$ mW, $V_{IN} = 1.9$ V, $V_{OUT} = 0.9$ V, and including the parasitic capacitances from the PMOS capacitor for each cell. The black solid line is the resulting frequency behaviour of the overall converter. The graphs show that the operation frequency is kept close to its maximum value and that the minimum frequency for each code decreases with the output power increase. Therefore, the cell controller must change the frequency limit depending on the code. The same frequency behaviour could have been obtained by only using unit cells all equal to C_{LSB}, requiring a total of seven cells, where an increase, or decrease, in the active cells number would increase, or decrease, P_{OUT} by P_{LSB}.

If the change in F_{CLK} for enabling or disabling one cell is a constant factor, the cell controller complexity can be decreased. Because now the cell controller only needs two limits to change the active cells. Therefore, a maximum and minimum frequency is defined as F_{max} and F_{min}, respectively. For this to be possible, the cells have to be enabled, or disabled, using a binary thermometer code, and so the cell's C_{FLY} size is calculated by determining the value of the output power at F_{min} when $C_{FLY} = C_{TOTAL}$. This gives the minimum output power value, P_{min}, at which the last cell is disabled. This output power value is then the maximum output power at F_{max}, when the last cell is disabled. Hence, the C_{FLY} cell value of the highest power cell is the difference between C_{TOTAL} and the required C_{FLY} value for the converter to deliver P_{min} at F_{max}. This repeats until the first cell is reached, which is simply equal to the second cell. This can be written in mathematical notation, by manipulating the C_{FLY} and P_{OUT} equations, resulting in the following equation:

$$C_{cell_i} = \begin{cases} \left(\dfrac{F_{min}}{F_{max}}\right)^{N-i} \left(1 - \dfrac{F_{min}}{F_{max}}\right) C_{FLY_{TOTAL}} & N \geq i > 1 \\[4mm] \left(\dfrac{F_{min}}{F_{max}}\right)^{N-1} C_{FLY_{TOTAL}} & i = 1 \end{cases} \tag{5.5}$$

where $i \in \mathbb{N}$ and N is the total cell number for the converter to be divided. It can be easily observed that if the relation of the maximum and minimum frequency is 1/2, then the equation simplifies to

$$C_{cell_i} = \begin{cases} \left(\dfrac{1}{2^{N-i}}\right) C_{FLY_{TOTAL}} & N \geq i > 1 \\[4mm] \left(\dfrac{1}{2^{N-1}}\right) C_{FLY_{TOTAL}} & i = 0 \end{cases} \tag{5.6}$$

which results in one plus $N-1$ binary-weighed cells that have a binary-weighed power behaviour. Figure 5.11 shows the plot of the F_{CLK} Eq. (5.2) with multiple cells (dashed lines) and different F_{min}/F_{max} relations, where each cell C_{FLY} is sized using (5.5), for $P_{OUT} = 16$ mW, $V_{IN} = 1.9$ V, $V_{OUT} = 0.9$ V, and the parasitics of the PMOS capacitor. The black solid line is the resulting frequency behaviour of the converter. It can be seen that, for the same number of cells, the closer F_{min} is to F_{max} the larger the power range

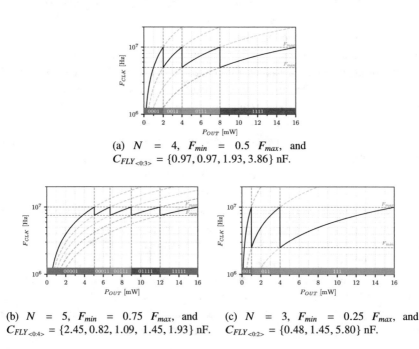

(a) $N = 4$, $F_{min} = 0.5\ F_{max}$, and $C_{FLY_{<0:3>}} = \{0.97, 0.97, 1.93, 3.86\}$ nF.

(b) $N = 5$, $F_{min} = 0.75\ F_{max}$, and $C_{FLY_{<0:4>}} = \{2.45, 0.82, 1.09,\ 1.45, 1.93\}$ nF.

(c) $N = 3$, $F_{min} = 0.25\ F_{max}$, and $C_{FLY_{<0:2>}} = \{0.48, 1.45, 5.80\}$ nF.

Fig. 5.11 Operation frequency range of the 1/2 CR SC converter with multiple cells and different F_{min}/F_{max} relations, for $P_{max} = 16$ mW, $F_{CLK_{max}} = 10$ MHz, $V_{IN} = 1.9$ V, $V_{OUT} = 0.9$ V, $\alpha = 3\%$, and $\beta = 0\%$

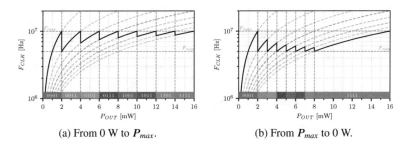

(a) From 0 W to P_{max}.   (b) From P_{max} to 0 W.

Fig. 5.12 Operation frequency range of the 1/2 CR SC converter with 1+3 binary-weighted cells behaviour for $P_{max} = 16$ mW, $F_{CLK_{max}} = 10$ MHz, $V_{IN} = 1.9$ V, $V_{OUT} = 0.9$ V, $\alpha = 3\%$, and $\beta = 0\%$

of operation of the first cell, and the farther F_{min} is from F_{max} then the last cells have the larger operation range. Thus, more cells are required if F_{min} is close to F_{max}. By using $F_{min} = 0.5\,F_{max}$, then, except for the first cell, all the capacitance values of the other cells are binary weighted, thus reducing the layout complexity, and enabling or disabling the cells' results in a binary-weighted output power. However, the step in the output power is not constant. Hence, the maximum power step will always be half the maximum power, and this can cause a considerable voltage ripple when toggling the last cell.

The two previous examples can be combined where the first cell is always active, $(1\times)$ C_{LSB}, and then N binary-weighted cells, $(1\times, 2\times, 4\times \ldots 2^N\times)$ C_{LSB}, are used to increase the effective capacitance in C_{LSB} increments, where the first cell is required to keep the output voltage constant when all the other cells are disabled. Figure 5.12a, b shows the example for $N = 4$, in the previous conditions as the graphs before. Now, if the cell controller only has two limits to determine when to change C_{FLY}, then this would result in a different behaviour when the power increases from 0 W to P_{max}, and when decreases from P_{max} to 0 W. This reduces the complexity of the cell converter even further, because it naturally adds hysteresis between the cell transitions, at the penalty of working at a lower maximum frequency when the output power is decreasing. Nonetheless, this maximum frequency when decreasing the power increases as the power decreases, which is in accordance with the increase in the output voltage ripple due to the power reduction.

The same analysis was made for the 2/3 and 1/1 converter, which can be seen in Fig. 5.13. This was done considering the capacitor sized for the 1/2 CR, as explained above. While the frequency behaviour of the 2/3 CR is similar to the 1/2 CR, the 1/1 CR requires a higher maximum frequency value than $F_{CLK_{max}}$ to reach the 16 mW. This means that the maximum frequency and output power will be limited by the 1/1 CR, thus the capacitor must be sized so that $F_{CLK_{max}}$ occurs at P_{max} in the CR 1/1. Meaning that the other two CRs will have the flying capacitance slightly oversized. Hence, by sensing the frequency it is possible to enable or disable the converter cells to reduce the flying capacitance size and thus reducing the output voltage ripple.

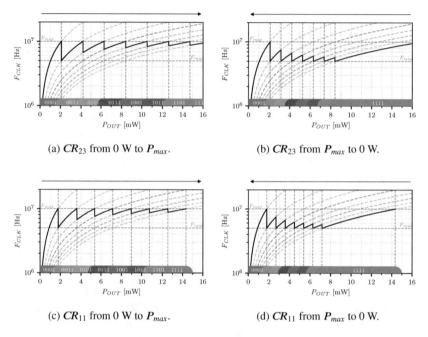

(a) CR_{23} from 0 W to P_{max}.

(b) CR_{23} from P_{max} to 0 W.

(c) CR_{11} from 0 W to P_{max}.

(d) CR_{11} from P_{max} to 0 W.

Fig. 5.13 Operation frequency range of the 2/3 and 1/1 CR SC converter with 1+3 binary-weighted cells behaviour, sized according with the CR_{12}, for $F_{CLK_{max}} = 10$ MHz, $V_{IN} = 1.5$ V for the CR_{23} and $V_{IN} = 1.1$ V for the CR_{11}, $V_{OUT} = 0.9$ V, $\alpha = 3\%$, and $\beta = 0\%$

The number of cells has to be chosen knowing that it is inversely proportional to the output voltage ripple (as shown in (2.42) of Sect. 2.4, Chap. 2). Therefore, each converter unit cell was divided into 32 smaller cells (slice unit cells). This number was obtained by running electrical transient simulations, of the circuit with different cell numbers, until the voltage ripple was within the accepted limits. To further reduce the voltage ripple, the disabled cells C_{FLY} are connected in parallel with the output node, thus increasing the total output decoupling capacitor. Furthermore, as explained later in this chapter, a large number of interleaved cells facilitate the cell controller operation. For 32 interleaved cells, the power per slice unit cell of cell 1 is $P_{OUT_{max}}/32 = 62.5$ μW. Therefore, the slice unit cell flying capacitor was sized so that the maximum frequency occurs at 16 mW in the 1/1 CR, which is the worst case. This results in a slice unit cell flying capacitor of 17 pF, corresponding to a total capacitance for the slice unit cell of 34 pF, to keep the maximum frequency at 10 MHz in the worst case. During the layout of the circuit, it was decided to increase the value of the slice unit cell flying capacitor to 20 pF, giving a total of 40 pF per cell. The interleaved cells of cell 2 are equal to cell 1, since $P_{OUT_{max}}$ is the same. As for the interleaved cells of cells 3 and 4, they were made by placing two interleaved cells of 62.5 μW in parallel (125 μW) for cell 2, and four interleaved cells of 62.5 μW in parallel (250 μW) for cell 4. Table 5.3 shows the value of each cell total flying capacitance and the maximum output power they

Table 5.3 Total flying capacitance per cell and its maximum output power

Cell #	C_{FLY} (nF)	P_{OUT} (mW)
1	1.28	2
2	1.28	2
3	2.56	4
4	5.12	8
Total	10.24	16

can give. Therefore, the total capacitance for the flying capacitance is 10.24 nF. The voltage limits are the same as in the previous section, which result in the maximum frequencies of 7.55 MHz for the 1/2 CR, 7.18 MHz for the 2/3 CR, and 8.43 MHz for the 1/1 CR. The switch's ON resistance sizing can be seen in Appendix A.

5.3 Auxiliary Circuits Design

The SC converter requires a set of auxiliary circuits to assure that it operates properly. This includes a phase generator to generate the time interleaved clock phases, implemented using an Asynchronous state machine (ASM); the loop regulation is carried out using a bandgap to generate the voltage reference, and a comparator to compare it with the divided output voltage; the active CR is set again by a dedicated ASM, where the divided input voltage is compared with the operation voltage limits of each CR; the number of active cells is set by another ASM, based on the clock frequency; each switch has a dedicated driver to ensure that it is driven properly, i.e. with a fast settling time and according with the active CR; finally, the start-up circuit to ensure the proper start-up of the circuit. The next sections describes each circuit in detail.

Almost all the circuits shown in the next sections have digital logic gates. Hence, before starting with the circuits description, a brief description of the logic gates will be shown here. Unless otherwise stated, all the logic gates used in this prototype are implemented with NOT, NAND, or NOR gates, whose circuit schematics can be seen in Fig. 5.14. All the other logic gates, e.g. AND, OR, or a Set-Reset (SR) latch, are made from these three logic gates, as shown in Fig. 5.15. The NMOS were sized with the minimum width that allows two contacts, 0.64/0.13 μm (W/L), and the PMOS transistors were sized with 2/0.13 μm (W/L) so that the ON/OFF transition is approximately at $V_{DD}/2$, where $V_{DD} = V_{OUT} = 0.9$ V.

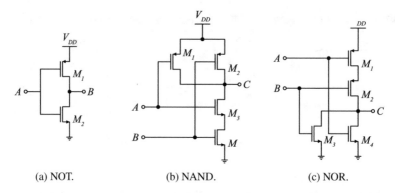

Fig. 5.14 Simplified schematic of the logic gates NOT, NAND, and NOR. PMOS and NMOS have their bulk connected to V_{DD} and ground, respectively

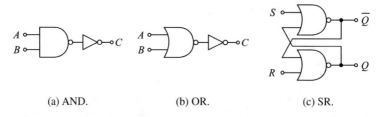

Fig. 5.15 Simplified schematic of the logic gates OR, AND, and SR

5.3.1 Phase Generator and Loop Regulation

The output voltage regulation is performed using frequency modulation, implemented through single-bound hysteretic regulation, where the block diagram is depicted in Fig. 5.16, which was explained in detail in Sect. 2.4 of Chap. 2. The working principle is as follows: a comparator compares the divided output voltage with an internal reference voltage and acts in an ASM that produces the clock signal. The ASM produces 32 clock signals, $\phi_{<0:31>}$, that are fed to a Non-overlap Clock Generator (NCG), that generates both ϕ_1 $_{<0:31>}$ and ϕ_2 $_{<0:31>}$. Moreover, due to the way the ASM is implemented, there is a pulse signal in each rising edge of ϕ_i. These pulse signals are combined in an OR logic gate to produce ϕ_p. This signal is used to sense the operation frequency of the converter as explained below.

Figure 5.17 shows the phase generator ASM simplified schematic. Assuming that V_C is high, the ASM works as follows: in the beginning V_{RESET} is high, and the ASM starts in the $\phi_{<0>}$ state. This signal goes through a delay circuit that delays it by t_d ($t_d = 1/(32 \, F_{CLK_{max}})$). The delayed version of $\phi_{<0>}$ is used to activate the next signal, $\phi_{<1>}$. This goes on until it reaches the signal $\phi_{<31>}$, thus completing the cycle, and returning to $\phi_{<0>}$. To ensure a duty cycle of 50%, the number of phases 32 was divided in half, which means that $\phi_{<16>}$ must go to logic low when $\phi_{<0>}$ goes to logic high, and this repeats throughout the clock

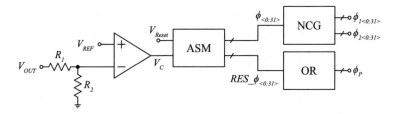

Fig. 5.16 Phase generator simplified block diagram

signals. This is shown in the ASM schematic, where the delayed version of $\phi_{<0>}$ resets the state that produces $\phi_{<17>}$, $\phi_{<1>}$ resets the state $\phi_{<18>}$, and so on. Extra logic gates were used to assure that the transitions between states are properly made. The comparator signal can stop the ASM in any of the clock signals, through an AND gate between the delayed version of ϕ and V_C. Hence, the ASM stops almost instantaneously (with a maximum time delay that corresponds to the delay between phases) when V_C is low. When V_C goes high again the ASM picks up where it has stopped. Figure 5.18 shows the transient simulation results of four clock signals ($\phi_{<0>}$, $\phi_{<1>}$, $\phi_{<16>}$, and $\phi_{<17>}$) of the PMU. It can be seen that $\phi_{<16>}$ is the complementary signal of $\phi_{<0>}$, and the same relation between $\phi_{<17>}$ and $\phi_{<1>}$.

Between each rising edge of the signals $\phi_{<0:31>}$ a reset signal is generated, $RES\_\phi_{0:31}$, that is used to force the transition from high to low of the corresponding phase. Hence, by combining these reset signals, a train of pulses is generated, where each pulse rising edge is evenly spaced by t_d, and thus 32 pulses are evenly spaced in a clock period. Therefore, each $\phi_{<0:31>}$ is high for 16 pulses and low for another 16 pulses. Since the PMU stops in the corresponding state whenever V_C is low, the total number of pulses gives the average frequency clock, which can be counted using a digital counter. Moreover, the detection method does not depend on the absolute frequency value nor the active CR, thus reducing the complexity of the cell controller. Figure 5.19a shows the simplified schematic of the OR and Fig. 5.19b the electrical simulation results of the described behaviour.

The delay of between each rising edge of $\phi_{<0:31>}$ is produced by an weak inverter delay, shown in Fig. 5.20. To generate the non-overlap ϕ_1 and ϕ_2 signals a conventional NCG was used, whose schematic can be seen in Fig. 5.21a. An extra AND was added with the signals ϕ_{1_END} and ϕ_{2_END} where the returning ϕ_1 and ϕ_2 signals from the cells drivers are connected. This assures that there is no overlap due to the clock lines path and drivers propagation delay. Each $\phi_{<0:31>}$ has a dedicated NCG. Figure 5.21b shows the schematic simulations of the circuit where the waveforms of $\phi_{1<0>}$ and $\phi_{2<0>}$ can be observed.

Figure 5.22 shows the schematic of the comparator circuit [28]. It is composed of a pre-amplification stage, composed of a differential pair with active loads. The differential pair connects to a positive feedback decision circuit, through M_{16} and M_{17} with their gates cross-connected, where M_{19} is used to move the decision circuit's output swing into the common-mode range of the next stage output buffer. A differential buffer is used to convert

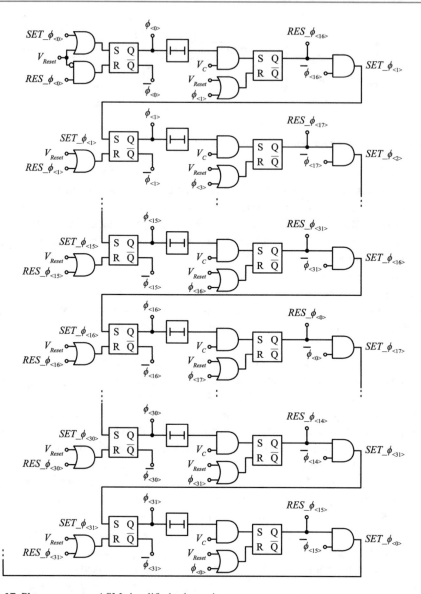

Fig. 5.17 Phase generator ASM simplified schematic

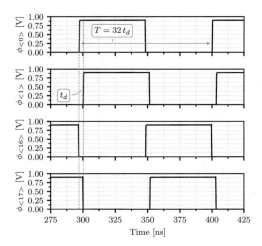

Fig. 5.18 Simulation results of the phase generator ASM $\phi_{<0:31>}$ signals

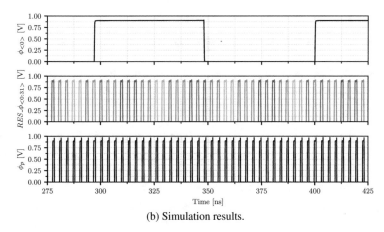

(a) Simplified schematic.

(b) Simulation results.

Fig. 5.19 Simplified schematic and simulation results of the pulse signal ϕ_p for frequency detection

Fig. 5.20 Simplified schematic
of the weak inverter delay

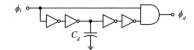

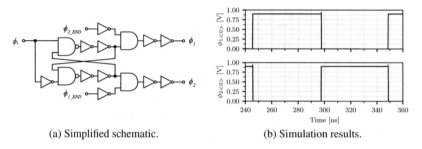

(a) Simplified schematic.   (b) Simulation results.

Fig. 5.21 Schematic and simulation results of the non-overlap clock generator

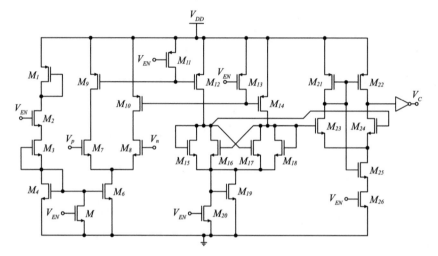

Fig. 5.22 Simplified schematic of the comparator. PMOS and NMOS have their bulk connected to V_{DD} and ground, respectively

the output of the decision circuit into a logic 0 or V_{DD} signal. Finally, an inverter is added to the output to isolate the comparator from any load capacitance [29]. The comparator can be powered down when not needed through the enable signal (V_{EN}).

Each converter cell has always 16 cells in ϕ_1 and other 16 cells in ϕ_2, and in ϕ_2, the flying capacitors are connected to the output node in all the CRs. Hence, the converter has at least always half of the active flying capacitance being used, connected to its output node. Moreover, since the disabled cells are connected in the phase ϕ_2 configuration, the decoupling capacitance is further increased with the cells being disabled. Hence, the worst case will be when all the cells are enabled, and thus the minimum value of the output decoupling capacitor would be half of the total flying capacitance, which gives 5.12 nF. Nonetheless, since local on-chip decoupling capacitors will be placed in the layout to occupy the empty spaces, this value will be further increased. Hence, the comparator response delay was determined using this capacitance value.

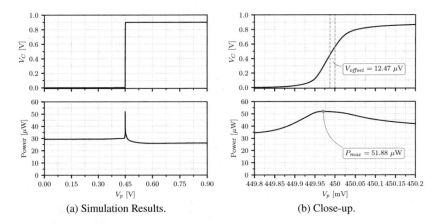

Fig. 5.23 Comparator DC simulation results with V_n=0.45 V, V_p swept from 0 to 0.9 V, and for $V_{DD} = 0.9$ V

Figure 5.23 shows the comparator's DC simulation results obtained by sweeping V_p from 0 to 0.9 V with $V_n = 0.45$ and $V_{DD} = 0.9$ V. These show the comparator's output voltage and power consumption. The systematic offset is equal to 12.47 μV and the power consumption is approximately equal for V_p at 0 V or 0.9 V, with the value of 29 μW and 26 μW, respectively. The maximum value is 52 μW.

The comparator's response delay will affect the minimum value of the converter's output voltage. Hence, to observe this, a simulation was performed, where a capacitor of 5.12 nF charged to 0.9 V (plus 20 mV to represent the ripple variation), was discharged by a 50.625 Ω resistor, resulting in a P_{OUT} of 16 mW. The capacitor's voltage was divided by 2, by a resistive divider and fed into V_n. The other input of the comparator, V_p, was fixed at 0.45 V and $V_{DD} = 0.9$ V. The comparator's output connects to a load capacitor of 80.2 fF, which is equal to the total gate capacitance of the logic gates connected at its output. Figure 5.24 shows the simulation results, where the delay in the comparator response is measured from the time instant when $V_p = V_n = 0.45$ V and the time instant when the comparator's output voltage reaches 90% of V_{DD}. This gives a delay of approximately 5.8 ns, which results in a V_{droop} on the capacitor of 8.8 mV, that results in 17.6 mV on the converter's output.

Table 5.4 shows the corner simulation for Typical-Typical (tt), Slow-Slow (ss), Slow NMOS-Fast PMOS (snfp), Fast NMOS-Slow PMOS (fnsp), $V_{DD_{nom}} \pm 10\%$, and for 0° to 100°, of the comparator power consumption and systematic offset. The table shows that in the worst case, the output voltage could drop approximately 34 mV. Monte Carlo simulation was performed (500 cases) for the tt corner, with $T = 300$ K and $V_{DD} = 0.9$ V, to determine the input referred offset, which resulted in a mean value of -0.84 mV with a worst-case deviation of 7.67 mV.

Fig. 5.24 Comparator transient simulation results with V_n connected to a resistive voltage divider that connects to a 5.12 nF capacitor and a 50.625 Ω resistor, with $V_p = 0.45$ V and $V_{DD} = 0.9$ V

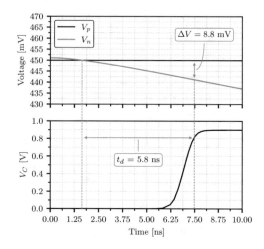

Table 5.4 Comparator corner simulation results for tt, ss, snfp, fnsp, $V_{DD_{nom}} \pm 10\%$ and for $0°$ to $100°$

Test	Nominal	Min	Max
V_{offset} (μV)	12.45	-7.039	47.63
P_{max} (μW)	51.87	22.47	96.42
t_d (ns)	5.83	4.19	10.59
ΔV (mV)	8.77	5.88	17.00

5.3.2   Voltage Reference Generator

To fully integrate the PMU a voltage reference for the loop regulation and the CR controller is required. Figure 5.25 shows the simplified schematic of the bandgap, including the two-state two-stage Miller compensated amplifier, chosen to generate a stable reference voltage V_{REF} of 450 mV [30].

This all CMOS bandgap voltage reference circuit [30] works as follows: the diode-connected transistors $M_{5,8}$ operate in the weak inversion region, where their current varies exponentially with $V_{GS_{5,8}}$. Thus, $V_{GS_{5,8}}$ is given by (5.7) where I_{D0} is the technology current, n is the substrate value, V_T is the thermal voltage, L and W are the transistor length and width, and V_{th} is the threshold voltage. For the 130 nm CMOS technology, the values of I_{D0} and n for an NMOS transistor were determined by simulation and are 0.9 μA, and 1.17, respectively.

$$V_{GS} = n \, V_T \, \ln\left(\frac{I_D \, L}{I_{D0} \, W}\right) + V_{th} \qquad (5.7)$$

The transistors $M_{5,8}$ along with the resistances $R_{1,2}$ are responsible for the generation of a Proportional to Absolute Temperature (PTAT) and Complementary to Absolute Temperature

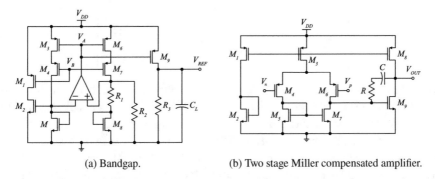

(a) Bandgap.   (b) Two stage Miller compensated amplifier.

Fig. 5.25 Simplified schematic of the voltage reference circuit. PMOS and NMOS have their bulk connected to V_{DD} and ground, respectively

(CTAT) current. The values of these currents can be derived by applying KCL to node V_p (5.8), where it is assumed that the amplifier gain is high enough so that $V_n = V_p$, and that the current I_D is equal in all the four branches of the circuit and so it can be derived through Ohm's law (5.9)

$$\frac{V_{gs5} - V_{gs8}}{R_1} + \frac{V_{gs5}}{R_2} - I_D = 0 \tag{5.8}$$

$$I_D = \frac{V_{REF}}{R_3} \tag{5.9}$$

Considering $L_5 = L_8 = L, V_{th5} = V_{th8}, I_{D05} = I_{D08} = I_{D0}, n_5 = n_8 = n, I_{D5} \approx I_{D8} \approx I_D$, and $W_5 = K\,W_8$, then $V_{gs5} - V_{gs8}$ from (5.8) can be replaced by (5.7), resulting in

$$V_{gs5} - V_{gs8} = n\,V_T \ln \left(\frac{\dfrac{I_D\,L}{I_{D0}\,W_5}}{\dfrac{I_D\,L}{I_{D0}\,K\,W_5}} \right) = nV_T \ln (K) \tag{5.10}$$

Replacing (5.10) in (5.8) and (5.9) results

$$V_{REF} = \frac{R_3}{R_2} \left(V_{gs5} + \frac{R_2}{R_1} n\,V_T \ln (K) \right) \tag{5.11}$$

Hence, the variation of V_{REF} over the temperature T is given by

$$\frac{\partial V_{REF}}{\partial T} = \frac{R_3}{R_2} \left(\frac{\partial V_{gs5}}{\partial T} + M \frac{\partial V_T}{\partial T} \right) \tag{5.12}$$

where $M = \frac{R_2}{R_1} n \ln (K)$. For V_{REF} to be constant its derivative must be equal to zero, thus

$$\frac{R_3}{R_2}\left(\frac{\partial V_{gs5}}{\partial T} + M\frac{\partial V_T}{\partial T}\right) = 0 \Rightarrow M = \left|\frac{\frac{\partial V_{gs5}}{\partial T}}{\frac{\partial V_T}{\partial T}}\right| \qquad (5.13)$$

where $\partial V_T/\partial T = k/q = 86.3\ \mu V/^{\circ}C$, in which k is the Boltzmann constant ($k = 1.38 \times 10^{-23}$ J/K) and q is the elementary charge ($q = 1.60 \times 10^{-19}$ C). By simulating an NMOS diode-connected transistor biased by a current source (Fig. 5.26a) it is possible to determine the V_{gs} variation with temperature (Fig. 5.26b). The current was set to 500 nA and $L = 10\ \mu m$, the temperature was swept from -40 to $100\,^{\circ}C$ for different W values (1 μm, 5 μm, 10 μm, and 20 μm). As expected from (5.7) as W decreases V_{gs} increases. Also, V_{gs} decreases with the increase in the temperature, i.e. it has a negative temperature coefficient like the bipolar transistors. For the chosen current and L values, and knowing that $V_{th} = 0.220$ V (extracted by simulation), then W must be higher than 5 μm to be in the moderate region ($V_{gs} \approx V_{th}$) or weak inversion ($V_{gs} < V_{th}$). Table 5.5 shows the slope, i.e. $\partial V_{gs}/\partial T$, for each W value from Fig. 5.26b. Using the value obtained from simulation for $W_5 = 5\ \mu m$ results in $M = 7.35$ (5.13).

The sizing of the resistors $R_{1,2,3}$ was performed using the same method used in [31], where the total resistance value was set according with the area constraints which resulted

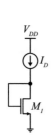

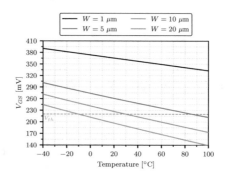

(a) Schematic.

(b) Simulation results of the V_{gs} as a function of the temperature for different W values, with $I_D = 500$ nA and $L = 10\ \mu m$.

Fig. 5.26 Simplified schematic and simulation results of the test bench used to determine the $\partial V_{gs5}/\partial T$ value. The transistor's bulk is connected to ground

Table 5.5 Simulation results of $\partial V_{gs}/\partial T$ of an NMOS transistor with $L = 10\ \mu m$ and $I_D = 500$ nA

W (μm)	$\partial V_{gs}/\partial T$ ($\mu V/^{\circ}C$)
1	-395.6
5	-634.5
10	-697.7
20	-750.4

in a total resistance (R_{Total}) value of 4 MΩ. Then, at room temperature ($T = 300$ K), V_{gs} ($W = 5$ μm) $= 258$ mV, $V_T = 0.026$ V, $V_{REF} = 0.45$ V, $K = 4$ (for employing common centroid techniques), then

$$V_{REF} = \frac{R_3}{R_2}\left(V_{gs5} + M\,V_T\right) \Longleftrightarrow 0.45 = \frac{R_3}{R_2}\,(0.258 + 7.35 \times 0.026) \tag{5.14}$$

$$M = \frac{R_2}{R_1}\,n\,\ln(K) \Longleftrightarrow 7.35 = \frac{R_2}{R_1} \times 1.17 \times \ln(4) \tag{5.15}$$

$$R_{total} = R_1 + R_2 + R_3 = 4 \times 10^6\ \Omega \tag{5.16}$$

Solving (5.14)–(5.16) results in

$$R_1 = 397\ K\Omega \tag{5.17}$$

$$R_2 = 1.80\ M\Omega \tag{5.18}$$

$$R_3 = 1.80\ M\Omega \tag{5.19}$$

These resistance values were used as a starting point and were then adjusted through electrical simulations until the minimum variation of V_{REF}, as a function of the temperature, was achieved. Moreover, the current on M_9 was increased by a factor of 4 and a Metal-Insulator-Metal (MIM) decoupling capacitor (C_L) of 20 pF was added so that the reference voltage is stable enough to withstand the kickback noise from the comparators, without suffering significant voltage fluctuations. Hence, R_3 was divided by 4, to keep V_{REF} at 450 mV. The final values obtained through simulation are

$$R_{1_{sim}} = 309.1 K\Omega \tag{5.20}$$

$$R_{2_{sim}} = 2.62\ M\Omega \tag{5.21}$$

$$R_{3_{sim}} = 350.7\ K\Omega \tag{5.22}$$

The bandgap is also responsible for generating the voltage limits of each converter CR. This is done by designing the bandgap output resistance R_3 using a resistive ladder. Hence, R_3 was divided into $R_{3a} = 39.40$ KΩ, $R_{3b} = 15.76$ KΩ, $R_{3c} = 47.28$ KΩ, $R_{3d} = 15.76$ KΩ, $R_{3e} = 232.46$ KΩ. These generate the voltage limits for the CR transitions scaled down by a factor of 5. Hence, the voltage limits are 1.9 and 1.5 V, on the transitions between the CRs 1/2 to 2/3 and 2/3 to 1/1. As for the transition limits between the CRs 1/1 to 2/3 and 2/3 to 1/2 are 1.6 and 2.0 V, which have 50 mV of hysteresis, from the previous limits, to avoid multiple successive transitions between CRs. The bandgap's resistive ladder is implemented using unit resistors to improve matching. Hence, its value can be adjusted by connecting or not connecting unit resistors in series. Therefore, the last five unit resistors plus five extra unit resistors of R_{3e} have an NMOS switch in parallel. These switches are controlled by a D flip-flop that enables or not the switch according to a digital signal. There are a total of 12 bits in a thermometer configuration which allows to trim the bandgap's output between 505 mV and 395 mV, with a step of approximately 10 mV.

Table 5.6 Simulation
results of the bandgap's
two-stage Miller
compensated amplifier

Parameter	Value
Gain	82 dB
GBW	13.4 MHz
Phase margin	90.6°

The two-stage Miller compensated amplifier is required to assure that $V_n = V_p$ so that the current in both branches is equal. The simulation results can be seen in Table 5.6. It has a DC gain of 82 dB, with a Gain-Bandwidth (GBW) of 13.37 MHz and a phase margin of 90.6°.

Figure 5.27 shows the simulation results for V_{REF} as a function of the temperature, for a V_{DD} of 0.9 V, and as a function of V_{DD} at 27 °C, respectively. The first (Fig. 5.27a) shows that the maximum and minimum values of V_{REF} are 450.91 mV and 449.94 mV, respectively. Resulting in a total variation of 0.96 mV, and a temperature coefficient ($TC = (\Delta V_{REF} \, 1 \times 10^6)/(V_{REF\ 27°C} \, \Delta T)$) of 15.32 ppm/°C. The second, Fig. 5.27b, shows that the bandgap voltage is constant at 0.45 V for V_{DD} values larger than 0.7 V. Figure 5.28 shows the Power Supply Rejection Ratio (PSRR) performance for $V_{DD} = 0.9$ V at 27 °C. The lowest PSRR value of 29 dB occurs for frequencies around 1 MHz.

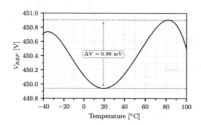

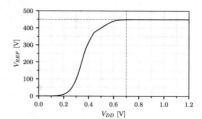

(a) V_{REF} as a function of the temperature for $V_{DD} = 0.9$ V.

(b) V_{REF} as a function of the supply voltage V_{DD} at 27°C.

Fig. 5.27 DC simulation results of the bandgap circuit

Fig. 5.28 Simulation results of
the PSRR of the bandgap
circuit for $V_{DD} = 0.9$ V and
$T = 27$°C

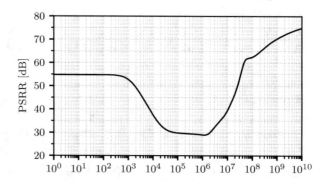

Table 5.7 Bandgap post-layout corner simulation results summary, for tt, ss, snfp, fnsp, max, min, and $V_{DD_{nom}} \pm 10\%$

Test	Nominal	Min	Max
$V_{REF\ 27\,°C}$ (mV)	450.0	403.4	505.8
ΔV_{REF} (mV)	0.96	0.87	28.0
TC ppm/°C	15.32	13.77	496.0

Table 5.7 shows the post-layout corner simulation results summary of the bandgap. A total of 65 simulations were performed for tt, ss, Fast-Fast (ff), snfp, fnsp, max and min for the resistance and capacitances, and finally $V_{DD} \pm 10\%$.

Figure 5.29 shows the simulation results of the bandgap reconfiguration for a $V_{DD} = 0.9$ V at 27 °C. The graph shows the calibration mechanism that allows to change the bandgap voltage between 405 mV and 505 mV in steps of approximately 10 mV, so that the voltage reference can be adjusted on the prototype measurements if needed.

Fig. 5.29 Simulation results of the bandgap reconfigurability, for $V_{DD} = 0.9$ V and $T = 27°C$

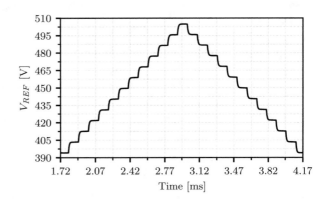

The bandgap start-up is assured by the circuit shown in Fig.5.30. It works by pulling down to 0 V the amplifier's output node (V_A) and the cascode node (V_B), allowing for the current mirror to start working and thus starting up the bandgap. This start-up mechanism is controlled using the V_{RESET} from the start-up circuit. Hence, when V_{RESET} is high, the transistors $M_{1,4}$ are turned ON and a small current, controlled by $M_{2,3}$ and $M_{5,6}$ (with large L values) flows to ground and thus forcing the nodes to 0 V. When V_{RESET} is low, $M_{1,4}$ are turned OFF and no current flows in $M_{2,3}$ and $M_{5,6}$. Hence, the bandgap start-up at the same time has the rest of the PMU.

Fig. 5.30 Simplified schematic of the bandgap circuit. PMOS and NMOS have their bulk connected to V_{DD} and ground, respectively

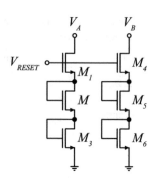

Table 5.8 Required voltages for each switch in each clock phase and CR

CR	1/2		2/3		1/1	
Phase	ϕ_1	ϕ_2	ϕ_1	ϕ_2	ϕ_1	ϕ_2
S_1 (PMOS)	V_{DDL}	V_{DDH}	V_{DDL}	V_{DDH}	V_{DDL}	V_{DDH}
S_2 (PMOS)	V_{DDL}	V_{DDH}	V_{DDL}	V_{DDH}	V_{DDL}	V_{DDH}
S_3 (PMOS)	V_{DDH}	0	V_{DDH}	0	V_{DDH}	0
S_4 (PMOS)	V_{DDH}	0	V_{DDH}	V_{DDH}	V_{DDH}	0
S_5 (NMOS)	0	V_{DDL}	0	0	V_{DDL}	V_{DDL}
S_6 (NMOS)	0	V_{DDL}	0	V_{DDL}	V_{DDL}	V_{DDL}
S_7 (PMOS)	0	V_{DDL}	0	V_{DDL}	V_{DDL}	V_{DDL}
S_8(PMOS)	0	V_{DDL}	0	V_{DDL}	V_{DDL}	V_{DDL}
S_9 (NMOS[a])	0	0	0	V_{DDH}	0	0

[a]3.3 V Device

5.3.3 Switches' Drivers

The voltage applied to each switch varies with both the selected CR and clock phase. Hence, different clock signals for each switch are required for the converter to work properly and to assure that the transistors' breakdown voltage is not exceeded. Table 5.8 shows the required voltages for each switch according to the CR and clock phase, where V_{DDH} is the highest voltage, in this case V_{IN}, and V_{DDL} is the lowest voltage, that ideally would be V_{OUT}. However, in some cases, it may be $V_{IN} - V_{MAX}$, to ensure that the switch's voltage swing is within its safe limits, ch5:

Has shown in Table 5.8, a clock swing from the highest voltage ($V_{DDH} = V_{IN}$) to 0 V is required in some switches. This was done using a conventional level shifter, shown in Fig. 5.31a. It works as follows: one of the NMOS ($M_{1,2}$) will be turned ON by ϕ_L creating a low impedance path at the drain node, and the other will be turned OFF, creating a high impedance path at the drain node. As the low impedance node discharges, one of the cross-coupled PMOS ($M_{3,4}$) will turn ON, and it will pull the floating node to V_{DDH}, while the

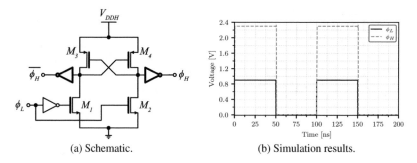

(a) Schematic.   (b) Simulation results.

Fig. 5.31 Simplified schematic of the conventional level shifter. The thick line indicates that 3.3 V transistors were used. PMOS and NMOS have their bulk connected to V_{DDH} and ground, respectively

other PMOS transistor will be turned OFF. This generates a clock signal with a swing of 0 V to V_{DDH}, and its complementary signal. This clock signal is required by the switches S_1, S_2, S_3, and S_4. Figure 5.31b shows the transient simulation results of the level shifter for $V_{DDH} = 2.3$ V, for a 10 MHz input clock signal with 0.9 V amplitude.

Each interleaved cell has its own drivers since each one works with a different delayed clock signal. The switch drivers S_1 and S_2 (Fig. 5.32a) are equal, and composed of the level shifter followed by an inverter that operates between V_{DDH} and V_L, where V_L is a voltage generated by a resistive divider that produces the necessary voltage for the switch V_{GS} to be within the maximum voltage limits. The same level shifter generates the clock signals for the switches S_5 and S_6, and each goes through a Multiplexer (MUX) implemented by transmission gates for generating a different clock signal on the CR_{11}. The switches S_3 and S_4 do not require a level-up signal, hence they are only controlled through two MUXs (Fig. 5.32b) that change the clock signals depending if it is on the CR_{23} or CR_{11}. Finally, the drivers of the switches S_7, S_8, and S_9 are implemented with a level shifter (Fig. 5.32c) and again using MUXs to change the clock signal on the CR_{23}. Since the switches controlled by this signal have a low width value, it was not required to add extra inverters to have a fast transition. The clock non-overlap is assured because the slowest switch clock signals generated from ϕ_1 ($V_{S1,S2}$) and from ϕ_2 (V_{S8}) are used in the NCG as the signals ϕ_{1_END} and ϕ_{2_END}.

5.3.4   Conversion Ratio Controller

The converter's current CR 1/2, 2/3, or 1/1 (which will be named CR_{12}, CR_{23}, and CR_{11}, respectively) is determined according with the V_{IN} value and the operation voltage limits of each CR. Figure 5.33 shows the CR controller block diagram [32, 33]. It is composed of four main blocks: a voltage limit selector, whose function is to generate the voltage limit to be compared with V_{IN} according with the current CR; a dynamic comparator, to produce a

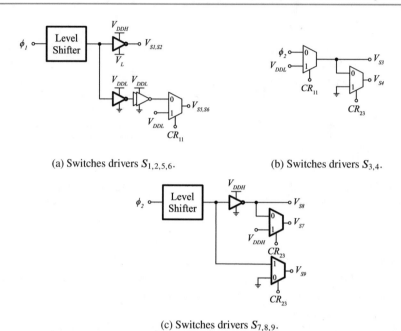

(a) Switches drivers $S_{1,2,5,6}$.   (b) Switches drivers $S_{3,4}$.

(c) Switches drivers $S_{7,8,9}$.

Fig. 5.32 Simplified schematic of switches' drivers. The thick line indicates 3.3 V transistors

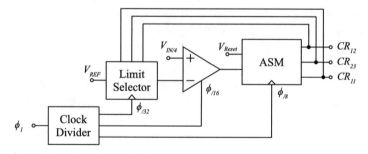

Fig. 5.33 Simplified block diagram of the CR controller

control signal when it is time to change the current CR; an ASM to generate the control signals for each CR; and finally a clock divider that generates three clock signals, from ϕ_1, divided by 8, 16, and 32. These are fed to the ASM, comparator, and limit selector, respectively, to decrease the dynamic power dissipation by reducing the number of comparisons per second.

Since the supercapacitor voltage changes slowly, it is not necessary to check every clock cycle if the CR must be changed. Therefore to reduce the power dissipation, ϕ_1 was divided by 8 (further dividing would not result in a significant power reduction), and two extra clocks were also generated by further dividing by two. Hence, three new clocks are generated ($\phi_{/8}$, $\phi_{/16}$, and $\phi_{/32}$) that are ϕ_1 divided by 8, 16, and 32, respectively. Figure 5.34a shows the circuit of the clock divider using a master-slave transmission gate D flip-flop scheme. It is composed of NOT gates and transmission gates. To generate $\phi_{/8}$, three dividers by two were cascaded, and then $\phi_{/16}$ is generated by dividing by two $\phi_{/8}$, and $\phi_{/32}$ is generated by dividing by two $\phi_{/16}$. The clock signal $\phi_{/8}$ is used to clock the ASM. The clock signal $\phi_{/16}$ is used to clock the comparator. And finally, $\phi_{/32}$ is used by the limit selector. This clock scheme was chosen to give enough time for the signals to settle before the comparator makes a decision. Figure 5.34b shows the simulation results of the clock divider. When $\phi_{/32}$ is low, the upper limit is connected to the comparator's input, and when high, the lower limit is connected to the comparator's input. The comparator checks the difference between the divided V_{IN} and the selected voltage limit when $\phi_{/16}$ is high, allowing for the voltage limit to settle during half a period of $\phi_{/16}$. The ASM only checks the comparator decision when $\phi_{/8}$ and $\phi_{/16}$ are high. Hence, it allows half of the $\phi_{/8}$ period for the comparator voltage to settle. To conclude, if $\phi_{/32}$ is low and $\phi_{/8}$ and $\phi_{/16}$ are high, the controller checks if it must change to an upper CR. If $\phi_{/32}$ is high and $\phi_{/8}$ and $\phi_{/16}$ are high, the controller checks if it must change to a lower CR.

Figure 5.35 shows the simplified schematic of the limit selector clock. It is composed of a resistive ladder (from the bandgap) that is used to generate the voltage limit levels for each CR, from V_{REF}. There are a total of four voltage limits, the transition between the CR_{11} and CR_{23} (V_a), and CR_{23} and CR_{11} (V_b), and to avoid multiple transitions between CRs around the voltage limits, extra voltages (V_{ah} and V_{bh}) spaced by 10 mV from the original voltage transitions were added, to create hysteresis. These voltage limits are selected through transmission and logic gates, according with the active CR and $\phi_{/32}$. Only CR_{23} will have a lower and upper limit, because there is no change below CR_{11} neither above CR_{12}.

Figure 5.36 shows the simulation results of the limit selector. The graph shows that on CR_{12} and CR_{11} the voltage limits are static, independent of $\phi_{/32}$, and are equal to 0.38 mV (1.9 V) and 0.31 mV (1.55 V), respectively. On the CR_{23} the voltage limit changes with $\phi_{/32}$ and is equal to 0.3 mV (1.5 V), for the lower limit, and 0.39 mV (1.95 V) for the upper limit. The values in brackets are to the voltage limit values multiplied by 5.

The output of the limit selector and the divided input voltage is compared by a conventional dynamic comparator [34, 35], shown in Fig. 5.37, which provides a high input impedance, rail-to-rail output voltage swing, and ideally no static power consumption. It works as follows: when $\phi_{/16}$ is low, the transistors $M_{8,9}$ pull both the comparator's outputs, V_{Op} and V_{On}, to V_{DD}, which is the starting condition for the comparison. This condition is called the reset phase. When $\phi_{/16}$ is high, the nodes V_{Op} and V_{On} start to discharge through the differential pair $M_{1,2}$ and M_3. They will discharge at different rates depending on the voltages at V_p and V_n. Assuming that $V_n > V_p$, V_{On} will discharge faster than V_{Op}. Hence, when V_{On} is below the $V_{DD} - |V_{thp}|$, M_7 will turn ON and initiate the latch regeneration caused by the back-to-back inverters $M_{4,5,6,7}$. Thus, V_{Op} is pulled to V_{DD} and V_{On} discharges to ground. An SR flip-flop is connected to the comparator's output to assure that the V_C is only high when $\phi_{/16}$ is high, that is, when a comparison is being made, and not due to the reset $\phi_{/16}$ is low. If $V_p > V_n$, the circuits work vice versa. Transistors $M_{10,11}$ are

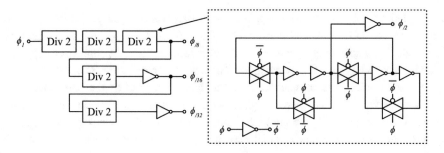

(a) Simplified schematic.

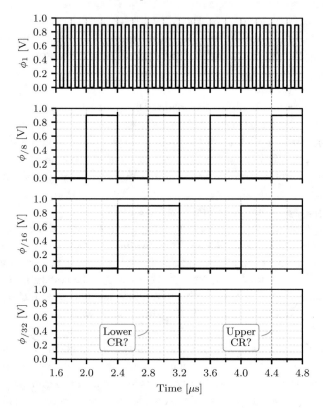

(b) Simulation results.

Fig. 5.34 Clock divider simplified schematic and simulation results

Fig. 5.35 Simplified schematic of the limit selector

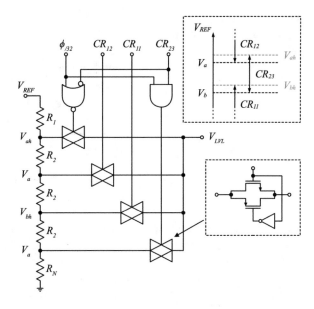

Fig. 5.36 Simulation results of the limit selector

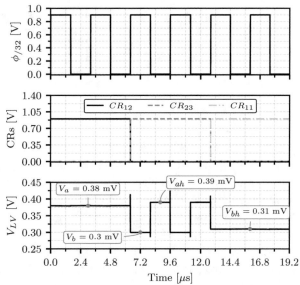

Fig. 5.37 Simplified schematic of the conventional dynamic comparator [34]

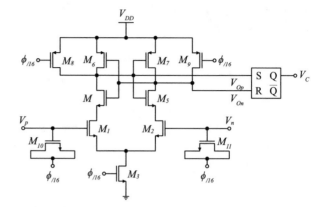

Fig. 5.38 Simulation results of the dynamic comparator in the comparison phase ($\phi_{/16} = V_{DD}$) for $V_{DD} = 0.9$ V, $V_n = 0.39$ V, and $V_p = V_n - 0.5$ mV

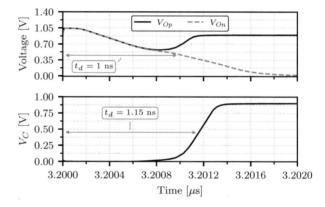

used to reduce the kickback noise. This noise comes from the leaking current caused by the parasitic capacitance between the drain and gate (C_{DG}) due to the large current flowing through $M_{1,2}$ when $\phi_{/16}$ is high. Thus, $M_{10,11}$, which are sized with half the size of $M_{1,2}$, act like a capacitor that injects a current that helps to cancel this leaking current [36].

Figure 5.38 shows the simulation results of the comparator, in the comparison phase ($\phi_{/16} = V_{DD}$), for $V_{DD} = 0.9$ V, $V_n = 0.39$ V, $V_p = V_n - 0.5$ mV. The behaviour described in the previous paragraph can be seen in the graph. It is defined as the converter delay time it takes for the comparator's output to reach $V_{DD}/2$. This resulted in a time delay t_d of 1 ns, for $V_{On} = 0.45$ V, and 1.15 ns for $V_C = 0.45$ V. Since the output of the comparator will only be checked 400 ns after the comparison (a pulse of $\phi_{/8}$), this value is more than enough for the comparator to decide. Table 5.9 shows the time delay for V_C for the four voltage limits. The worst case is 2.52 ns.

Figure 5.39 shows the simplified schematic of the CR controller ASM. It is composed of five states of operation that produce the three logic signals for each CR. It is assumed that the converter wakes up with the supercapacitor charging. Hence, it must initiate in the CR_{11}. To this end, the V_{RESET} signal ensures that the ASM starts on the state 1 (and all

Table 5.9 Dynamic comparator's V_C time delay t_d for different V_n levels, with $V_{DD} = 0.9\ V$ and $V_p = V_n - 0.5$ mV

V_n (V)	0.3	0.31	0.38	0.39
$V_C\ t_d$ (ns)	2.52	2.2	1.24	1.15

the other states are zero) which corresponds to the CR_{11}. Extra states, states 2 and 4, were added to ensure that the evaluation of a change in CR is not biased by the previous change conditions. Table 5.10 shows the lookup table of the auxiliary logic circuits A, B, and C. The CR is changed up when $\phi_{/16}$, $\phi_{/8}$ are high, and $\phi_{/32}$ and V_C are low. And changes down when $\phi_{/8}$, $\phi_{/16}$, $\phi_{/32}$, and V_C high.

Figure 5.40 shows the simulation results of the ASM, where it can be verified the transition between CRs following the rules of the lookup Table 5.9.

5.3.5 Cell Controller

Has explained in Sect. 5.2.4, the PMU's output power can be sensed through the clock frequency. Also, it has been as shown in the previous section, a pulse in every rise edge of $\phi_{<0:31>}$ is produced by the phase generator. Since the converter cells have 32 phases interleaved, and considering that V_{IN} is at the CR voltage limit value, it means that by counting the number of pulses the average frequency value can be determined (5.23), with a resolution of $1/32\ F_{max}$. If 32 pulses are counted, then the converter is working at maximum frequency, if only 16 are counted, it means it is at half the frequency, and so on. Hence, the cell controller does not need to know the absolute maximum frequency value, nor requires an always ON clock signal, it only needs to count the pulses from the phase generator while V_C is high. Therefore, it is basically composed of two counters, one to manage the cells and the other to count the number of pulses.

$$F_{CLK} = \frac{\text{Number of pulses}}{\text{Total pulses number}} \times F_{max} \tag{5.23}$$

Figure 5.41 shows an example of the explained method for sensing the clock frequency applied to a total of 4 clock pulses, instead of the 32 for simplicity. If 4 pulses are detected while V_C is high, then the converter is working at maximum frequency. If only 3 pulses are detected, then it is working at 3/4 of the frequency, and so on.

Has shown in Sect. 5.2.4 when using the $1 + 3$ binary weighted, the cell counter would increment the cell's code when the converter reaches the maximum frequency and it would decrement when the converter reaches the minimum frequency. Figure 5.42 shows the same plots, as shown in Fig. 5.12, with the number of pulses on the transition between cells. For example, when power is increasing. the transition between 1101 and 1111 brings the pulses

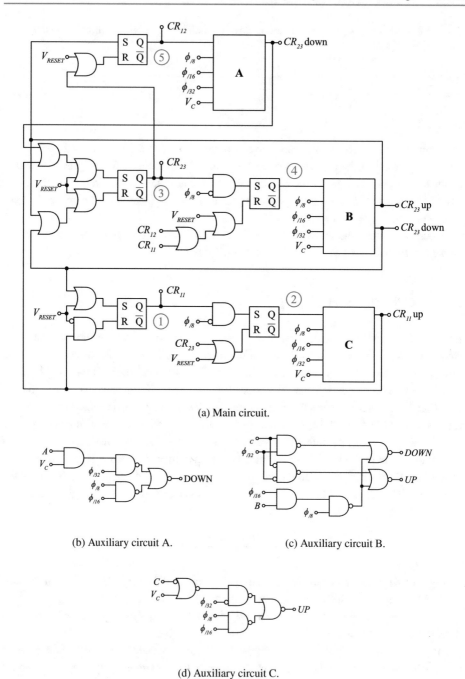

(a) Main circuit.

(b) Auxiliary circuit A.   (c) Auxiliary circuit B.

(d) Auxiliary circuit C.

Fig. 5.39 Simplified schematic of the CR controller ASM

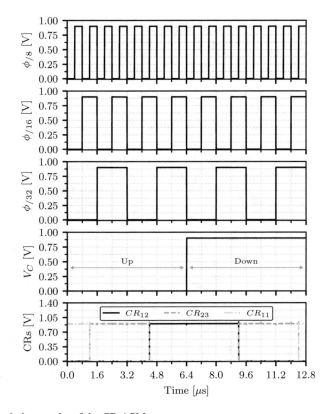

Fig. 5.40 Simulation results of the CR ASM

Table 5.10 ASM auxiliary circuits lookup table from the CR controller

$\phi/8$	$\phi/16$	$\phi/32$	V_C	Up	Down
1	1	1	1	0	1
1	1	0	0	1	0

down from 32 to 28. As for when power is decreasing, the transition between 1111 and 1101 increases the pulse count from 16 to 18.

This was the initial idea for the cell controller implementation. However, during the design process, it was noticed that when a full load step was applied, the output voltage would significantly drop. This is because the cell controller would only increase the number of cells after the 32nd pulse. Hence, when a transition between a no load condition to $P_{OUT_{max}}$ occurs, it would take 256 pulses to enable all cells, which since each phase (ϕ) period has 32 pulses, it would take 8 clock periods. This would result in a large voltage droop at the output, especially in this case, where no external output decoupling capacitor

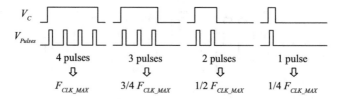

Fig. 5.41 Diagram of the F_{CLK} counter method for a total pulse number of 4

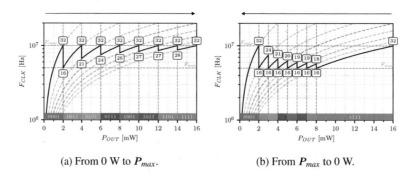

(a) From 0 W to P_{max}.   (b) From P_{max} to 0 W.

Fig. 5.42 Operation frequency range, with pulse information, of the 1/2 CR SC converter with 1+3 binary-weighted cell's behaviour for $P_{out_{max}} = 16$ mW, $F_{CLK_{max}} = 10$ MHz, $V_{IN} = 1.9$ V, $V_{OUT} = 0.9$ V, $\alpha = 3\%$, and $\beta = 0\%$

was used. The way this was fixed was to lower the F_{max} threshold (F_U) and dynamically decrease the pulse threshold limit number for enabling the next cell. This works as follows, starting on code 0001, when V_C goes high, the pulse counter starts to count. When 16 pulses are counted, the next cell is enabled (0011), and the pulse counter is reset. If V_C does not go low on the next 8 pulses, then the next cell is enabled (0101), and the pulse counter reset. While V_C is kept high, the pulse number threshold limit decreases to 4 pulses, and then to 2 pulses, where it stays until all cells are enabled. This way the converter responds to a load step in 36 pulses, i.e. approximately one period. This however brings the penalty of also lowering the lower frequency threshold (F_L) limit, and thus the converter now works close to half the maximum frequency. Nonetheless, this still results in a reduction of the output voltage ripple, when compared to all the cells enabled. Figure 5.43 shows the new operation frequency curve considering the F_U and F_L limits.

The output power is not the only variable that changes the frequency value, depending on the input voltage, the frequency will also change its value. Thus, it is only correct to say that the converter is working at the maximum power if the input voltage is at the CR voltage limit. Nonetheless, what will happen is that as the input voltage gets far from the CR voltage limit, the average frequency value will decrease, decreasing the number of cells. And, since the output voltage ripple increases with the input voltage being far from the CR voltage limit, the reduction of the cells will contribute to decrease the voltage ripple, by

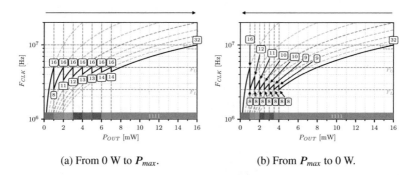

(a) From 0 W to P_{max}.   (b) From P_{max} to 0 W.

Fig. 5.43 Operation frequency range, with pulse information, of the 1/2 CR SC converter with 1+3 binary-weighted cell's behaviour for $P_{OUT_{aux}} = 16$ mW, $F_{CLK_{max}} = 10$ MHz, $V_{IN} = 1.9$ V, $V_{OUT} = 0.9$ V, $\alpha = 3\%$, and $\beta = 0\%$

adding more decoupling capacitance to the output. Hence, the voltage ripple reduction still holds, the cell control is actually two dimensional, where the number of active cells vary according to the output power and input voltage values.

Figure 5.44a shows the simplified schematic of the T flip-flop used in the cell counter. It is composed of two SR flip-flops in a master/slave configuration with a preset and clear signals, to bring Q high or low, respectively, instantaneously and independently of the counter bits or CLK signals. Figure 5.44b shows the simplified schematic of the 3-bit counter used to enable and disable the converter's cells. Where the signal down/$\overline{\text{Up}}$ controls if the cell's code increases (down/$\overline{\text{Up}} = 0$) or decreases (down/$\overline{\text{Up}} = V_{DD}$). The CLK signal is not going to be used as a clocked signal but rather as a pulse signal that goes high whenever a change in the counter is required, making this an asynchronous counter.

A 5-bit synchronous counter (Fig. 5.45) is used to count the number of pulses, pulse counter, while V_C is high. Therefore, the decision to enabling or disabling a cell is based on the output of this counter. The control circuit for these two counters is implemented using an ASM. Where its behaviour is as follows: when V_C goes high, the pulse counter starts to counter the number of pulses. If V_C goes low and 7 pulses are counted (7 is chosen instead of 8, to add one pulse of hysteresis to avoid multiple transitions in the transition between the codes 0001 to 0011), then the signal DEC goes high, and the cell counter is decreased by one. Hence, the cell's code decreases by one. If the number of counted pulses is between 8 and 15, then the signal SKIP goes high, and the cell controller does nothing. If the pulse counter reaches 16 pulses, then the signal INC goes high, and the cell counter is increased by one, increasing effective flying capacitance. The pulse counter is reset, to start a new count, and the decreasing cell mechanism goes through a cool down period, so that the cell's code is no mistakenly decreased if V_C goes down in less than 8 pulses. From now on, if V_C is kept high, the cell counter code is increased by one on the count of 8, 4, and 2 pulses, where the 2 pulses threshold is used to increase the cell counter until the maximum code is reached, i.e. all cells are activated. There are also two other control signals, one that goes high on the

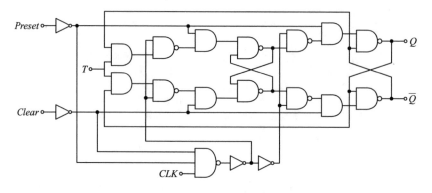

(a) T flip-flop schematic.

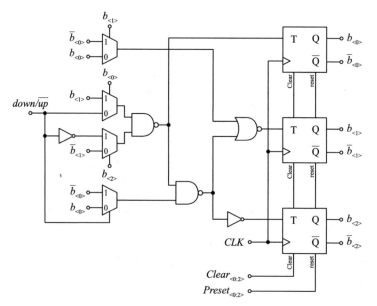

(b) Counter schematic.

Fig. 5.44 Simplified schematic of the 3-bit counter used to enable and disable the converter cells

0001 code and the other on the 1111 code, to prevent the ASM from going lower than 0001 and higher than 1111. The circuit control is implemented with logic gates and SR latches, which are not shown here due to the large circuit size.

Figure 5.46a shows the simulation results of the cell controller with V_C high long enough to make a transition between the code 0001–1111. The graph shows the increasing pulse threshold mechanism, where an INC signal goes high on the count of 16, 8, 2, and every 2 pulses. Because the pulse counter is reset when the pulse threshold limit is reached, and

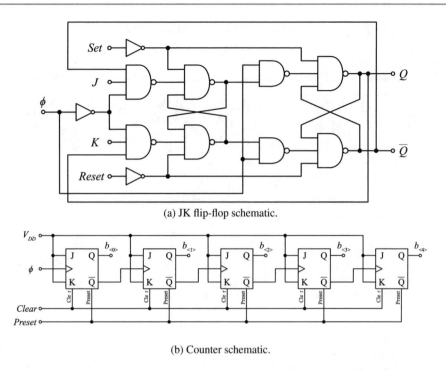

(a) JK flip-flop schematic.

(b) Counter schematic.

Fig. 5.45 Simplified schematic of the 5-bit counter used to count the pulse number

because there is a delay between INC signal and the change in the cell counter, some pulses are skipped between counts. Thus, instead of the 36 pulses, it takes 44 pulses, which means a time response of less than a clock period and a half. Figure 5.46b shows the transition between the codes 1111 to 1101. Only 4 pulses are counted (less than 8) and does DEC signal is triggered, which decrements the cell counter. In the next V_C pulse, 10 pulses are counted, hence there is not any action of increment or decrement, and so the SKIP signal goes high, so that the evaluation of increment or decrement in the cell counter is skipped.

Finally, special care must be taken on the CR transitions. When the transition is between $C R_{12}$ and $C R_{23}$, or $C R_{23}$ and $C R_{11}$, which means to change from a high-frequency to a low-frequency operation point, the pulse counter is reset, so a new evaluation can start without being biased from the number of pulses counted in the previous CR. For the transitions between a low frequency to high frequency of operation, i.e. $C R_{23}$ and $C R_{12}$ or $C R_{11}$ and $C R_{23}$, the cell activation mechanism is fast enough to ensure that the voltage drop at the output is acceptable.

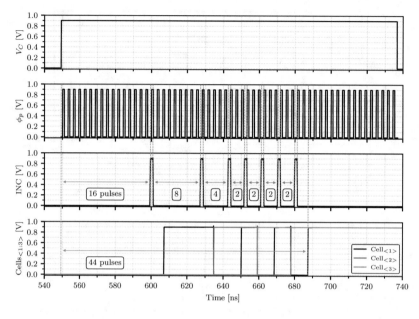

(a) From code 0001 to 1111.

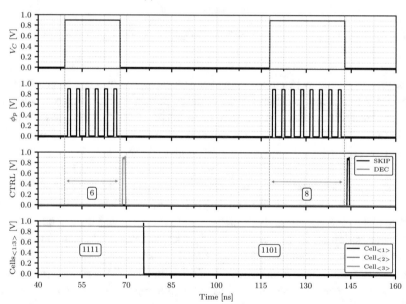

(b) From code 1111 to 1110.

Fig. 5.46 Simulation results of the cell controller transitions

5.3.6 Start-Up Circuit

The start-up circuit (Fig. 5.47) performs the circuit's start-up and generates a reset (V_{RESET}) and brown-out (V_{BROWN}) signals, used by the PMU circuits. It works as follows: M_1 turns ON when $V_{OUT} = 0$ V and $V_{IN} > V_{th1}$ connecting nodes V_{IN} and V_{OUT}, allowing V_{OUT} to start increasing with V_{IN}. V_1 is a scaled-down version by R_1 and R_2 of V_{OUT} that when reaches the threshold voltage of the first inverter, V_2 starts to decrease, V_3 to increase and V_{RESET} to decrease. A positive feedback provided by M_6 accelerates this transition by pulling up V_1, creating a fast rising edge. This causes the latch formed by M_2, M_3, M_4, and M_5 to connect V_{IN} to the gate of M_1, thus turning it OFF and ending the start-up process. The V_{BROWN} signal is produced when $V_{OUT} < 0.75$ V, by comparing V_{OUT} scaled by R_3 and R_4 with $V_{REF} = 0.45$ V. V_{BROWN} turns ON M_7 that pulls down V_1 (M_7 is sized to be three times larger than M_6) causing the start-up procedure to repeat.

Figure 5.48 shows the transient response of the start-up circuit with a ramp signal applied at the input voltage and with a capacitor of the same size as the total converter flying capacitance connected at the output of the start-up circuit. The graphs show the reset and brown-out mechanisms explained above. Table 5.11 shows the corner simulations for tt, ss, snfp, fnsp, max, min, 0 °C and 100 °C for only start-up circuit. It can be seen the variation of V_{IN} and V_{OUT} value when the reset and brown-out mechanism occurs, respectively. The minimum variation of the start-up input voltage does not go lower than 0.9 V, hence the PMU is able to start-up at the worst case. The V_{OUT} voltage for which the brown-out reset signal goes high does not have a significant variation.

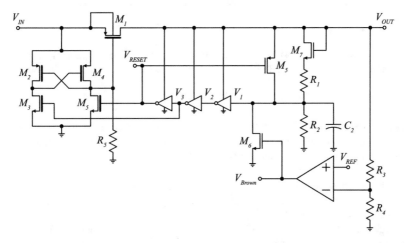

Fig. 5.47 Simplified schematic of the start-up circuit. The thick lines indicate 3.3 V devices. PMOS and NMOS have their bulk connected to V_{OUT}/V_{IN} and ground, respectively

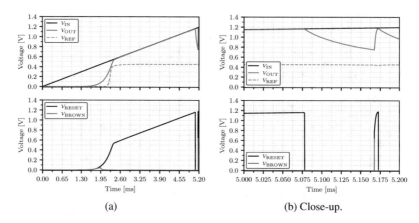

(a)   (b) Close-up.

Fig. 5.48 Simulation results of the start-up and brown-out mechanisms

Table 5.11 Start-up circuit corner simulation result summary for tt, ss, snfp, fnsp, resistors and capacitors max, min, 0 °C and 100 °C

Test	Nominal	Min	Max
V_{IN} when V_{RESET} goes low (V)	1.16	0.90	1.43
V_{OUT} when V_{BROWN} goes high (mV)	750.0	749.8	750.0

References

1. Harjani R, Chaubey S (2014) A unified framework for capacitive series-parallel DC-DC converter design. In: Proceedings of the IEEE 2014 custom integrated circuits conference. https://doi.org/10.1109/CICC.2014.6946050

2. Breussegem TV, Steyaert M (2013) CMOS integrated capacitive DC-DC converters. Springer, New York

3. Jiang J, Lu Y, Huang C, Ki W, Mok PKT (2015) A 2-/3-phase fully integrated switched-capacitor DC-DC converter in bulk CMOS for energy-efficient digital circuits with 14% efficiency improvement. In: 2015 IEEE international solid-state circuits conference (ISSCC). https://doi.org/10.1109/ISSCC.2015.7063078

4. Meyvaert H, Breussegem TV, Steyaert M (2011) A monolithic 0.77W/mm² power dense capacitive DC-DC step-down converter in 90nm Bulk CMOS. In: European solid-state circuits conference (ESSCIRC). https://doi.org/10.1109/ESSCIRC.2011.6045012

5. Ramadass YK, Chandrakasan AP (2007) Voltage scalable switched capacitor DC-DC converter for ultra-low-power on-chip applications. IEEE Power Electron Spec Conf. https://doi.org/10.1109/PESC.2007.4342378

6. Bang S, Seo J, Chang L, Blaauw D, Sylvester D (2016) A low ripple switched-capacitor voltage regulator using flying capacitance dithering. IEEE J Solid-State Circuits. https://doi.org/10.1109/JSSC.2015.2507361

7. Butzen N, Steyaert M (2016) Scalable parasitic charge redistribution: design of high-efficiency fully integrated switched-capacitor DC-DC converters. IEEE J Solid-State Circuits. https://doi.org/10.1109/JSSC.2016.2608349
8. Stathis JH (2002) Reliability limits for the gate insulator in CMOS technology. IBM J Res Dev. https://doi.org/10.1147/rd.462.0265
9. El-Damak D, Bandyopadhyay S, Chandrakasan AP (2013) A 93% efficiency reconfigurable switched-capacitor DC-DC converter using on-chip ferroelectric capacitors. In: IEEE international solid-state circuits conference digest of technical papers. https://doi.org/10.1109/ISSCC.2013.6487776
10. Andersen TM, Krismer F, Kolar JW, Toifl T, Menolfi C, Kull L, Morf T, Kossel M, Brändli M, Buchmann P, Francese PA (2015) A feedforward controlled on-chip switched-capacitor voltage regulator delivering 10W in 32nm SOI CMOS. In: IEEE international solid-state circuits conference (ISSCC). https://doi.org/10.1109/ISSCC.2015.7063076
11. Kudva SS, Harjani R (2013) Fully integrated capacitive DC-DC converter with all-digital ripple mitigation technique. IEEE J Solid-State Circuits. https://doi.org/10.1109/JSSC.2013.2259044
12. Le H, Sanders SR, Alon E (2011) Design techniques for fully integrated switched-capacitor DC-DC converters. IEEE J Solid-State Circuits. https://doi.org/10.1109/JSSC.2011.2159054
13. Le H, Seeman M, Sanders SR, Sathe V, Naffziger S, Alon E (2010) A 32nm fully integrated reconfigurable switched-capacitor DC-DC converter delivering 0.55W/mm^2 at 81% efficiency. In: IEEE international solid-state circuits conference - (ISSCC). https://doi.org/10.1109/ISSCC.2010.5433981
14. Breussegem TMV, Steyaert MSJ (2011) Monolithic capacitive DC-DC converter with single boundary-multiphase control and voltage domain stacking in 90 nm CMOS. IEEE J Solid-State Circuits. https://doi.org/10.1109/JSSC.2011.2144350
15. Piqué GV (2021) A 41-phase switched-capacitor power converter with 3.8 mV output ripple and 81% efficiency in baseline 90 nm CMOS. In: IEEE international solid-state circuits conference. https://doi.org/10.1109/ISSCC.2012.6176892
16. Le H, Crossley J, Sanders SR, Alon E (2013) A sub-ns response fully integrated battery-connected switched-capacitor voltage regulator delivering 0.19W/mm^2 at 73% efficiency. In: IEEE international solid-state circuits conference (ISSCC). https://doi.org/10.1109/ISSCC.2013.6487775
17. Jain R, Geuskens B, Khellah M, Kim S, Kulkarni J, Tschanz J, De V (2013) A 0.45–1V fully integrated reconfigurable switched capacitor step-down DC-DC converter with high density MIM capacitor in 22 nm tri-gate CMOS. In: IEEE symposium on VLSI circuits
18. Andersen TM, Krismer F, Kolar JW, Toifl T, Menolfi C, Kull L, Morf T, Kossel M, Brändli M, Buchmann P, Francese PA (2014). In: IEEE international solid-state circuits conference digest of technical papers (ISSCC). https://doi.org/10.1109/ISSCC.2014.6757351
19. Sarafianos A, Steyaert M (2015) Fully integrated wide input voltage range capacitive DC-DC converters: the folding Dickson converter. IEEE J Solid-State Circuits. https://doi.org/10.1109/JSSC.2015.2410800
20. Biswas A, Sinangil Y, Chandrakasan AP (2015) A 28 nm FDSOI integrated reconfigurable switched-capacitor based step-up DC-DC converter with 88% peak efficiency. IEEE J Solid-State Circuits. https://doi.org/10.1109/JSSC.2015.2416315
21. Tsai J, Ko S, Wang C, Yen Y, Wang H, Huang P, Lan P, Shen M (2015) A 1 V input, 3 V-to-6 V output, 58%-efficient integrated charge pump with a hybrid topology for area reduction and an improved efficiency by using parasitics. IEEE J Solid-State Circuits. https://doi.org/10.1109/JSSC.2015.2465853
22. Lu Y, Jiang J, Ki W (2017) A multiphase switched-capacitor DC-DC converter ring with fast transient response and small ripple. IEEE J Solid-State Circuits. https://doi.org/10.1109/JSSC.2016.2617315

23. Jiang Y, Law M, Chen Z, Mak P, Martins RP (2019) Algebraic series-parallel-based switched-capacitor DC-DC boost converter with wide input voltage range and enhanced power density. IEEE J Solid-State Circuits. https://doi.org/10.1109/JSSC.2019.2935556
24. Breussegem TV, Steyaert M (2010) A fully integrated gearbox capacitive DC/DC-converter in 90nm CMOS: Optimization, control and measurements. In: IEEE 12th workshop on control and modeling for power electronics (COMPEL). https://doi.org/10.1109/COMPEL.2010.5562379
25. Ramadass YK, Fayed AA, Chandrakasan AP (2010) A fully-integrated switched-capacitor step-down DC-DC converter with digital capacitance modulation in 45 nm CMOS. IEEE J Solid-State Circuits. https://doi.org/10.1109/JSSC.2010.2076550
26. Saurabh C, Harjani R (2017) Fully tunable software defined DC-DC converter with 3000X output current 4X output voltage ranges. In: IEEE custom integrated circuits conference (CICC). https://doi.org/10.1109/CICC.2017.7993625
27. Butzen N, Steyaert MSJ (2017) Design of soft-charging switched-capacitor DC-DC converters using stage outphasing and multiphase soft-charging. IEEE J Solid-State Circuits. https://doi.org/10.1109/JSSC.2017.2733539
28. Carvalho C, Lavareda G, Lameiro J, Paulino N (2011) A step-up μ-power converter for solar energy harvesting applications, using Hill Climbing maximum power point tracking. In: IEEE international symposium of circuits and systems (ISCAS). https://doi.org/10.1109/ISCAS.2011.5937965
29. Baker RJ (2010) CMOS circuit design, layout, and simulation. Wiley-IEEE Press
30. Quendera F, Paulino N (2015) A low voltage low power temperature sensor using a 2nd order delta-sigma modulator. In: Conference on design of circuits and integrated systems (DCIS). https://doi.org/10.1109/DCIS.2015.7388608
31. Pereira MS, Costa JEN, Santos M, Vaz JC (2015) A 1.1 μA voltage reference circuit with high PSRR and temperature compensation. In: Conference on design of circuits and integrated systems (DCIS). https://doi.org/10.1109/DCIS.2015.7388564
32. Madeira R, Paulino N (2016) Analysis and implementation of a power management unit with a multiratio switched capacitor DC-DC converter for a supercapacitor power supply. Int J Circuit Theory Appl. https://doi.org/10.1002/cta.2209
33. Madeira R, Oliveira JP, Paulino N (2018) A 130 nm CMOS power management unit with a multi-ratio core SC DC-DC converter for a supercapacitor power supply. IEEE Trans Circuits Syst II: Express Briefs. https://doi.org/10.1109/TCSII.2018.2861082
34. Kobayashi T, Nogami K, Shirotori T, Fujimoto Y (1993) A current-controlled latch sense amplifier and a static power-saving input buffer for low-power architecture. IEEE J Solid-State Circuits 10(1109/4):210039
35. Babayan-Mashhadi S, Lotfi R (2014) Analysis and design of a low-voltage low-power double-tail comparator. IEEE Trans Very Large Scale Integr (VLSI) Syst. https://doi.org/10.1109/TVLSI.2013.2241799
36. Eichenberger C, Guggenbuhl W (1989) Dummy transistor compensation of analog MOS switches. IEEE J Solid-State Circuits 10(1109/4):34103

Fully Integrated Power Management Unit Layout, Simulation and Measurements Results

6

6.1 Overview

This chapter covers the implementation of the 16 mW Power Management Unit (PMU) Integrated Circuit (IC) prototype, in 130 nm Complementary Metal-Oxide-Semiconductor (CMOS) bulk technology, to evaluate its performance and to validate the theoretical equations and schematic simulations. Hence, the following sections describe the layout design process in detailed, validated using electrical simulations. The Printed Circuit Board (PCB) test board and the test setup used for the circuit evaluation are then described. And finally, the measurement results of the prototype are presented, compared with the simulation ones, and conclusions are drawn.

6.2 Power Management Unit Layout and Floorplan

Figure 6.1a shows the PMU layout and floorplan, which was included in a 5×5 mm die together with other circuits. The PMU was divided into 3 main blocks—The converter cells, and digital and analogue blocks. Where the digital and analogue blocks were separated to try to minimize the coupling effect from the digital circuitry. Furthermore, both were surrounded by n-well guard rings to improve the circuit isolation from substrate noise. MOS decoupling capacitors were placed on empty spaces whenever possible. The total active area is 3.2×1.6 mm $= 5.12$ mm^2. An I/O pad ring composed of a set of IP cells,[1] with Electrostatic discharge (ESD) protection, was used in the die to provide connections to the circuits' inputs and outputs. The PMU uses a total of 6 pads: V_{IN}, V_{OUT}, V_{REF}, V_{PD}, and V_{SS}; and 8 other pads were used for an on-chip shift register, that acquires the Conversion

---

[1] Faraday Technology Corporation.

© The Author(s), under exclusive license to Springer Nature Switzerland AG 2022
R. Madeira et al., *Fully Integrated Switched-Capacitor PMU for IoT Nodes*, Synthesis Lectures on Engineering, Science, and Technology,
https://doi.org/10.1007/978-3-031-14701-2_6

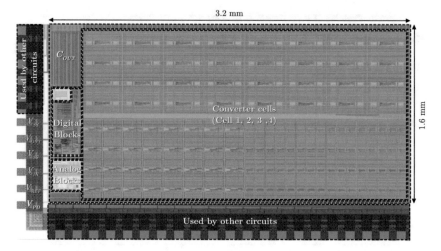

(a) PMU layout and floorpan.

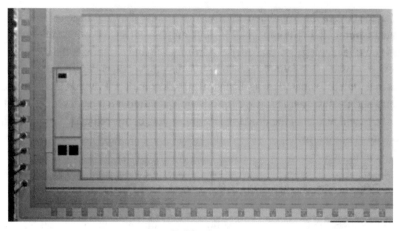

(b) Die photo.

Fig. 6.1 PMU 16 mW layout and floorpan and die photo

Ratio (CR) and cell signals, and sets the voltage reference calibration. The V_{REF} pad was intended for measurements purposes, and V_{PD} was used to power done the PMU when other circuits in the die are being tested. The development and implementation of the shift register are out of the scope of this work. An Field Programmable Gate Array (FPGA) Carrier Module (CMOD) S6 is used to communicate with the shift register allowing to change the reference voltage value and also to see the cells and CR signals state. Figure 6.1b shows the die photograph with the wire bonding.

6.2.1   Converter Cell Layout

Figure 6.2a shows the four main cells, where each one has 32 interleaved unit cells. These have an area of 2.9×1.57 mm^2, which occupies the majority of the total PMU area. The cells were made with the same height and the width proportional to the cell's power. Hence, the cell 1 and 2 widths are equal, cell 3 width is the sum of cells 1 and 2 width, and finally cell 4 width is two times cell 3 width, or four times cells 1 or 2 width. This was done so that they could be placed in a rectangular shape, has shown in the figure, where the 32 phase clock signals run through the middle and are distributed between the cells, in alternated ME4 and ME5 metals.

Figure 6.2 shows the unit interleaved cells layout of each 32 interleaved main cell. The switches, and their respective drivers, were placed between the two flying capacitors and the local decoupling capacitors (C_{OUT}) were placed around the cell. Figure 6.2b and 6.2c, show the layout of the 62.5 μW unit cell, used to form the 32 interleaved cells, of the 2 mW cells 1 and 2, respectively. The only difference between them is that cell 2 has two extra MUXs to control the cell's activation. The total size of the cell is 183×90 μm. Cell 3 unit cell, shown in Fig. 6.2d, was designed for a power of 125 μW, and its flying capacitance is made by placing two flying capacitors in parallel to double the total flying capacitance. The unit cell has the size of 183×180 μm, which means that it has the double of the cell 1 and cell 2 width. Finally, the cell 4 unit cell, shown in Fig. 6.2e, was designed to have a power of 250 μW, and its flying capacitance is made by placing four flying capacitors in parallel to quadruplicate the total flying capacitance. The unit cell has the size of 183×360 μm, which means that it has the quadruple of cells 1 or 2 width. The signals V_{IN} (ME8), V_{OUT} (ME7), and V_{SS} (ME6) were laid in vertical and horizontal lines, in a matrix configuration, on top of the cell, this way these signals are easily connected between the capacitors, drivers, and switches, and it also facilitates the connection between cells. Furthermore, this helps to reduce the line resistance due to the long connection.

6.2.2   Digital Blocks Layout

The layout view of the digital blocks can be seen in Fig. 6.3a and has a total area of 604×212 μm. The phases generator is shown in Fig 6.3b, with an area of approximately 212×501 μm. The 32 clock generator Asynchronous state machine (ASM) was designed so that each row of logic gates corresponds to a clock phase ($\phi_{<0:31>}$) that is shared vertically through a bus at the right side, containing the 32 clock phases, which were laid out perpendicularly to the ASM in alternating ME3 and ME5 lines. The logic signals to interconnect the rows were laid out in the same manner, in the left side of the ASM. N-well and P-sub connections were laid out between each logic gate row, to try to minimize the noise between each row, and also, to easily share V_{DD} and V_{SS} between them. The non-overlap clock phase generator, which generates the $\phi_{1,2}$ signals, which are fed to the converter cells, was laid

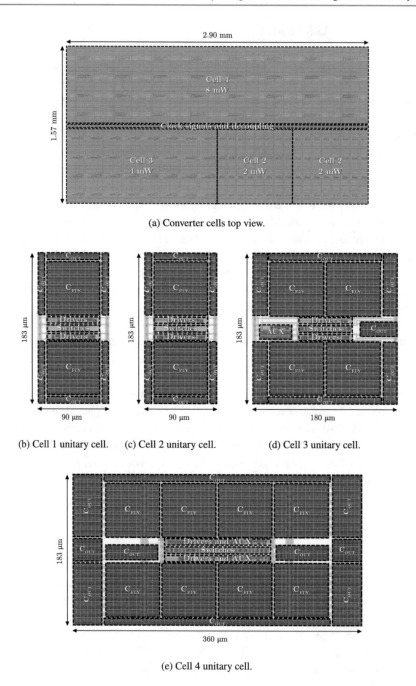

(a) Converter cells top view.

(b) Cell 1 unitary cell.   (c) Cell 2 unitary cell.   (d) Cell 3 unitary cell.

(e) Cell 4 unitary cell.

Fig. 6.2 Layout of the converter cells (2, 2, 4, and 8 mW) and the interleaved unitary cell of each main cell

out also in a row manner. Where the phases $\phi_{1,2}$ are grouped at the middle, to go between the converter's cell 4 and cells 1, 2, and 3. Starting from the top, the first two rows produce $\phi_{1<31>}$ and $\phi_{2<31>}$, the next two rows produce $\phi_{1<30>}$ and $\phi_{2<30>}$, and so on. Meaning that the clock phases bus at the middle has alternating ϕ_{1_i} (ME4) and ϕ_{2_i} (ME6) lines, that start at the top with the 32nd phase and goes all the way down until de 1st phase. Finally, the pulses of the ASM are combined using the OR block at the bottom of the ASM. Its layout was done in the same manner as the ASM. Decoupling capacitors were placed whenever there was space available.

Figure 6.3c and d show the cell and CR controller layout, respectively. They were laid out with the same design considerations described in the phase generator layout. The CR_{23} signal had to be up-converted to V_{IN} for the converter drivers, hence a level shifter was added to the CR controller.

6.2.3 Analogue Blocks Layout

Figure 6.4a shows the voltage reference generator and phase generator loop comparator layout. The design of the comparator, bandgap, and amplifier were made as symmetrical as possible, for each block. Dummy transistors were added to the boundary transistors to improve the matching. The resistors were laid out in an interdigitated manner to improve matching. As in the digital circuits, the analogue blocks, i.e. the voltage reference generator and the comparator, were surrounded by guard rings to improve the circuit isolation from substrate's noise. The start-up circuit layout can be seen in Fig. 6.4b, it has a total area of $100 \times 98\ \mu$m. The same layout design considerations as described above were used.

6.2.4 Simulation Results

Figure 6.5 shows the post-layout simulation results of the individual unit cells' efficiency and normalized drivers' power consumption (P_D/P_{OUT}), as a function of V_{IN}, were the frequency was calculated to make $V_{OUT} = 0.9$ V, with the cells connected to a load resistor (R_L) sized for their maximum power, i.e. $R_{Cell\#1} = R_{Cell\#2} = 12.96\,\mathrm{k}\Omega$, $R_{Cell\#3} = 6.48\,\mathrm{k}\Omega$, and $R_{Cell\#4} = 3.24\,\mathrm{k}\Omega$. For simulation purposes, each cell had a decoupling capacitor (C_L) of 100 times their flying capacitor connected to the output, to minimize the voltage ripple. The graph shows that the efficiency peak is higher in the cells 3 (85.0%) and 4 (85.6%), than in cells 1 (82.1%) and 2 (83.1%), due the higher P_{OUT} value, when compared to P_D. The dashed line shows the weighted average efficiency, which is similar to the cells 3 and 4 efficiency, and has a peak value of 84.6%. The normalized P_D values, in respect to P_{OUT}, increase as V_{IN} gets close to the CRs voltage limits. A maximum value of 12.5% in cells 1 and 2 is observed, and a maximum of 6.6 and 3.7% in cells 3 and 4, respectively.

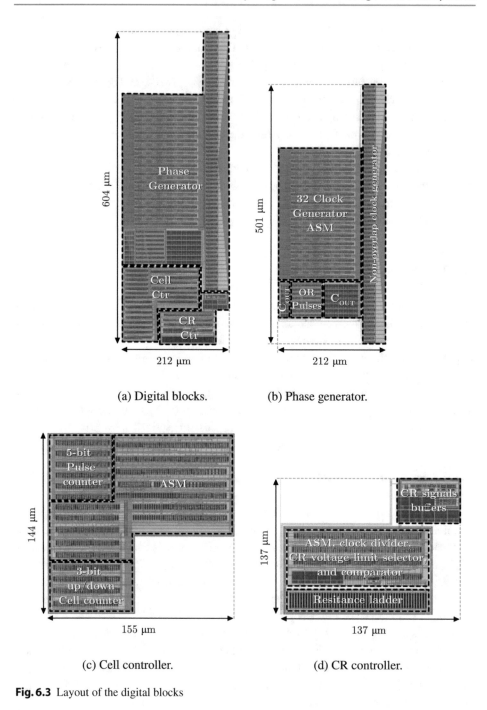

(a) Digital blocks.   (b) Phase generator.

(c) Cell controller.   (d) CR controller.

Fig. 6.3 Layout of the digital blocks

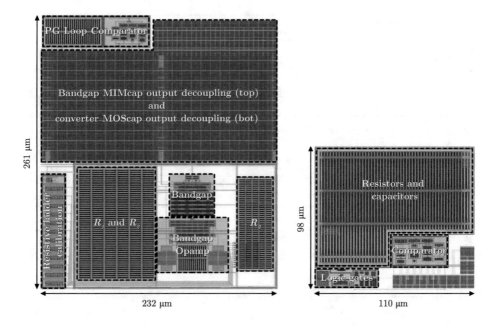

(a) Voltage reference generator and comparator.

(b) Start-up circuit.

Fig. 6.4 Layout of the analogue blocks

Fig. 6.5 Post-layout simulation results: Unit cells efficiency and normalized driver's power consumption (P_D/P_{OUT}) as a function of V_{IN}, for $V_{OUT} = 0.9$ V, $C_L = 100 \times C_{FLY}$, and $R_{Cell\#1} = R_{Cell\#2} = 12.96$ kΩ, $R_{Cell\#3} = 6.48$ kΩ, and $R_{Cell\#4} = 3.24$ kΩ

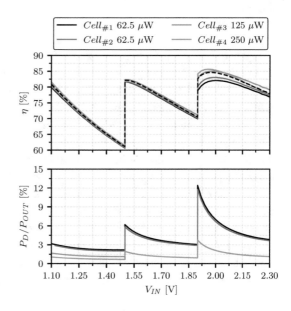

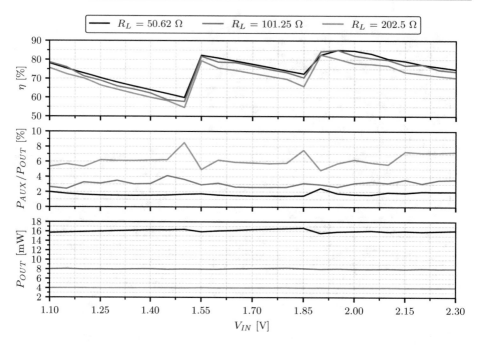

Fig. 6.6 Schematic simulation results: PMU overall efficiency as a function of V_{IN}, for $V_{OUT} = 0.9$ V and three different load resistor (R_L) values: 50.625 Ω (16 mW), 101.25 Ω (8 mW), and 202.5 Ω (4 mW)

Figure 6.6 shows the PMU's overall efficiency, normalized auxiliary circuits power (P_{AUX}/P_{OUT}), and output power, with the output voltage regulated to 0.9 V, and for three different load resistor (R_L) values: 50.625 Ω (16 mW), 101.25 Ω (8 mW), and 202.5 Ω (4 mW). The values shown in the graph were determined using electrical transient simulation[2] where a ramp from 0 to 2.3 V was applied to the PMU's input, for the PMU to reach to the CR_{12}, and then the input voltage ramps down to the V_{IN} value being simulated. After the PMU settle, i.e. changed the CR and the number of active cells if necessary, the input and output power are integrated for a long period of time, and thus the efficiency is obtained through P_{OUT}/P_{IN}. The maximum efficiency is approximately 85.1%, in the 1/2 CR, and the average efficiency is approximately 75.3%. This value is in good agreement with the previous efficiency values of the unit cells. Thus, close to the maximum output power (16 mW), the auxiliary circuits power has little impact on the overall efficiency, which is around 2% as shown in the graph. As expected, as the power decreases, the normalized auxiliary circuits' power consumption starts to dominate. For $R_L = 101.25$ Ω (8 mW) it increases to 3 to 4%

---

[2] Due to the circuit's complexity, it was not possible to simulate the extracted layout of the circuit. Hence, only schematic electrical simulations are shown form now on using the Spectre Accelerated Parallel Simulator (APS), to accelerate the transient electrical simulations.

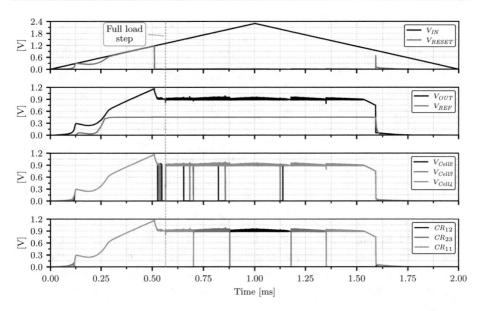

Fig. 6.7 Schematic simulation results: PMU transient response with a ramp signal from 0 to 2.3 V, and back to 0 V, applied to V_{IN}. At $V_{IN} = 1.3$ V a load of 50.625 Ω (16 mW) is connected to the output

of the output power, and for $R_L = 202.5$ Ω (4 mW) it further increases to values of 6 to 8%. This translates to a decrease in the efficiency values as the output power decreases.

Figure 6.7 shows the schematic simulation of the PMU with a ramp signal from 0 to 2.3 V, and back to 0 V, applied to V_{IN}. A 50.625 Ω load ($P_{OUT} = 16$ mW at $V_{OUT} = 0.9$ V) is connected when $V_{IN} = 1.3$ V and it remains connected until the end of the simulation. The graph shows the input ramp and the V_{RESET} signal, where it can be seen that starts the PMU when $V_{IN} \approx 1.2$ V. The second graph shows the V_{OUT} behaviour, which will be analysed in detail in the next graphs. Nevertheless, it can be seen that it remains fairly constant at 0.9 V. Despite the voltage droops on the CRs transitions, when V_{IN} is decreasing from 2.3 V, the maximum voltage ripple variation is approximately 71.2 mV, and a mean value of 39.6 mV. When V_{IN} is increasing towards 2.3 V, the maximum voltage ripple variation is approximately 78.4 mV, and a mean of 47.7 mV. This graph also shows the reference voltage V_{REF} which starts producing a stable voltage when V_{OUT} is close to 0.6 V. It then remains stable, regardless of the variation in V_{OUT}. The third graph shows the cells signals, which indicate if a given cell is active or not. In this case, the change in the cells is caused by the V_{IN} value, since P_{OUT} is constant. Finally, the fourth graph shows the CR signals that change according to V_{IN}.

As described above, at $V_{IN} = 1.3$ V a load of 50.625 Ω (16 mW) is connected at the output. Figure 6.8 shows a zoom in of the previous graph of the moment when the load is connected. Before the step, cells 2, 3, and 4 were disabled, and V_{OUT} is slowly discharging

Fig. 6.8 Schematic simulation results: PMU transient response to a full load step $R_L = 50.625\ \Omega$ (0–16 mW)

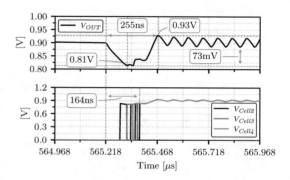

through the PMU's idle power consumption. When the load is connected, V_{OUT} starts to drop rapidly. It took approximately 70 ns for the PMU to respond to this event by enabling cell 2 (V_{Cell2} high). Approximately 90 ns later, the PMU has enabled all its cells, thus it took a total of 164 ns to respond to the load step. The output voltage stops decreasing when cell 2 is enabled, causing a voltage droop of 73 mV, and it took a total of 255 ns to reach the steady state.

Figures 6.9 show a zoom in on the transitions between the CRs 1/2 and 2/3. In this case, no significant voltage droop is observed. This can be explained since in both CRs, and in both phases, the flying capacitors are connected to the output node. On the other hand, the transition between the CRs 2/3 and 1/1 goes from a CR with the flying capacitor connected in both phases to the output, to one that only has the flying capacitors connected to the output in phase ϕ_2. Hence, as shown in Fig. 6.10, a voltage droop is observed, especially in the transition 2/3–1/1. In this transition, V_{OUT} has dropped approximately 90.3 mV and it took 183 ns to recover. In the transition between 1/1 and 2/3, V_{OUT} has dropped 25.3 mV and it took 331 ns to recover.

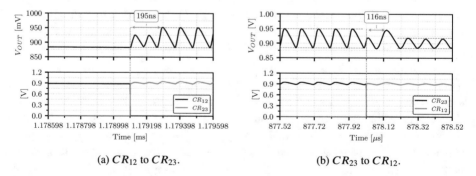

(a) CR_{12} to CR_{23}.   (b) CR_{23} to CR_{12}.

Fig. 6.9 Schematic simulation results: PMU transitions between CR 1/2 and CR 2/3 transient response, with $R_L = 50.625\ \Omega$ (16 mW)

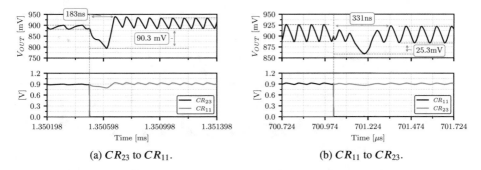

(a) CR_{23} to CR_{11}.   (b) CR_{11} to CR_{23}.

Fig. 6.10 Schematic simulation results: PMU transitions between CR 2/3 and CR 1/1 transient response, with $R_L = 50.625\ \Omega$ (16 mW)

Figure 6.11 show a zoom in of the output voltage and the reference voltage, with a load resistor of 50.625 Ω (16 mW), and with a ramp in the input voltage from 0 to 2.3 V and back to 0 V. It can be seen that the variations of the output voltage are attenuated in the reference voltage, which shows a maximum voltage variation 0.71% of its nominal value (0.45 V).

The same simulation setup used in Fig. 6.7 was simulated using different load resistor (R_L) values. The objective was to compare the difference between the output voltage ripple with and without the cell controller. Figure 6.12 shows the PMU transient response for an $R_L = 101.25\ \Omega$ (8 mW). In the V_{OUT} graph, the black line is the PMU with the cell controller and the grey line is the PMU without the cell controller, i.e. with all cells always enabled. With the cell controller, the voltage ripple is always below the value without the cell controller. However, there are some voltage droops or spikes caused by the cell activation not being instantaneous. Figures 6.13 and 6.14 show the same simulation for other load values:

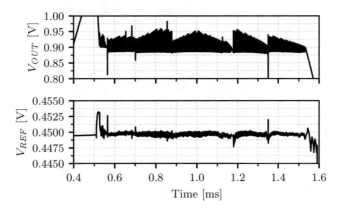

Fig. 6.11 Schematic simulation results: PMU reference voltage transient response to an input voltage ramp from 0 to 2.3 V and back to 0 V, with a load resistor of 50.625 Ω (16 mW)

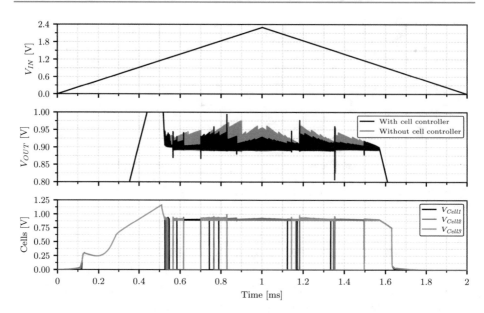

Fig. 6.12 Schematic simulation results: PMU transient response to an input voltage ramp signal from 0 to 2.3 V and back to 0 V, with $R_L = 101.25\,\Omega$ (8 mW)

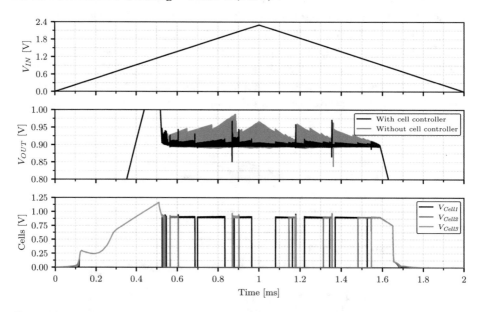

Fig. 6.13 Schematic simulation results: PMU transient response to an input voltage ramp signal from 0 to 2.3 V and back to 0 V, with $R_L = 202.5\,\Omega$ (4 mW)

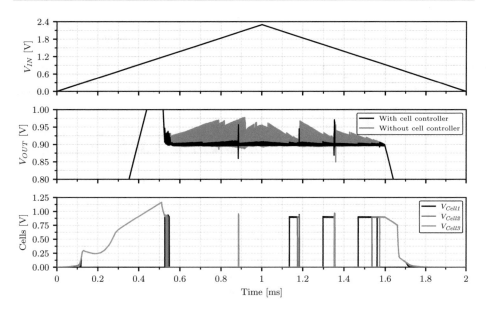

Fig. 6.14 Schematic simulation results: PMU transient response to an input voltage ramp signal from 0 to 2.3 V and back to 0 V, with $R_L = 405\ \Omega$ (2 mW)

$R_L = 202.5\ \Omega$ (4 mW) and $R_L = 405\ \Omega$ (2 mW), respectively. At these power levels, the decrease in the output voltage ripple is even more noticeable, than without having the cell controller.

The following graphs show the PMU response to a ramp in the output power. The PMU was started with a fixed input voltage and with an ideal current source connected to output, which was swept from 0 A (0 W) to the maximum output current value 17.78 mA (16 mW at 0.9 V) and back to 0 A (0 W). Figures 6.15 and 6.16 show the PMU transient electrical simulation results to the ramp in the output current for $V_{IN} = 1.95$ V (to be close to the 1/2 CR voltage limit). In Fig. 6.15 the vertical lines are placed at every rising edge of each cell, hence they represent a change in the active cell's number. The data tips show the output power value where the cell's code changed. For example, the first cell was activated at 1.4 mW, the second at 3.5 mW, and so on. Multiple cell activations are observed for the same power value. This can be explained by a voltage droop upon the cell activation, which triggers an increase in the cell's code to bring the output voltage back to its nominal value. This is what happens at 6.7 mW, where the cells code jumps from 4 to 6. The last cell is activated at 11.9 mW. The power values for which the cell's code increases (or decreases) depend also on the input voltage value. The closest the input voltage is to the CR voltage limits, the closest the transition will be to their designed value. Nevertheless, there will always be a ripple reduction with the cell controller, as long as the cells are not all active for the entire range. As expected, the change in the cells occurs at a lower power level when the power

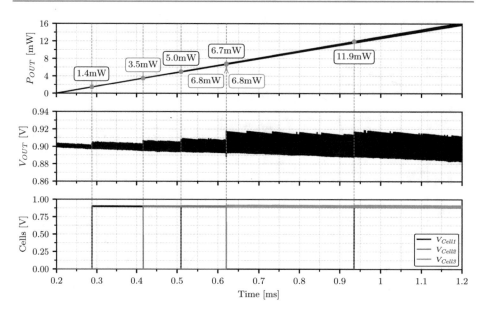

Fig. 6.15 Schematic simulation results: PMU transient response to an output current ramp signal from 0 A to 17.78 mA (16 mW at 0.9 V), with $V_{IN} = 1.95$ V

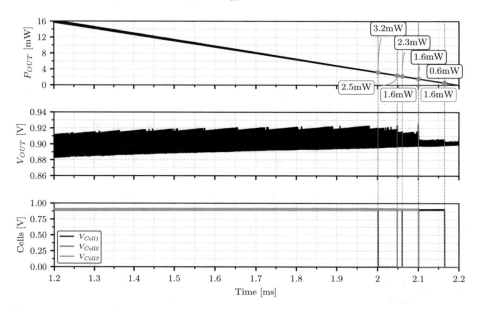

Fig. 6.16 Schematic simulation results: PMU transient response to an output current ramp signal from 17.78 mA (16 mW) to 0 A, with $V_{IN} = 1.95$ V

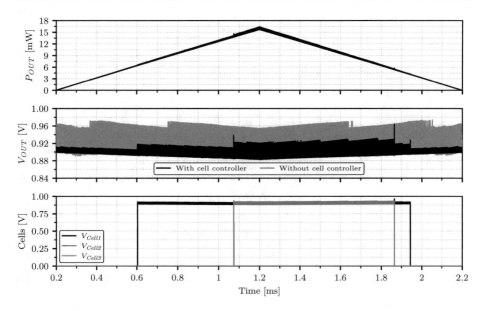

Fig. 6.17 Schematic simulation results: PMU transient response to an output current ramp signal from 0 A to 17.78 mA (16 mW) and back to 0 A, with $V_{IN} = 2.3$ V

is decreasing (Fig. 6.16). Now the vertical lines are placed on the falling edge of the cells, showing when the number of cells is decreased.

Figures 6.17, 6.18, 6.19, 6.20, and 6.21 show the same simulation performed in Figs. 6.15 and 6.16, for different V_{IN} values. These values were chosen to be close to the highest and lowest voltage of the CR limits. For example, for the 1/2 CR the voltages chosen were 2.3 V (maximum) and 1.95 V (close to the 1.9 V limit). The same was performed for the CR 2/3 and 1/1. It can be seen in all of them that having the cell controller substantially decreases the output voltage ripple.

Figure 6.22 shows a zoom in of the output voltage ripple comparison with and without the cell controller for different input voltage and output power values. It can be seen that the ripple can be reduced by more than 80% using the cell controller. Finally, Table 6.1 summarizes the performance of the proposed PMU.

6.2.5   Test Board and Measurements Setup

A PCB was designed using the EAGLE[3] layout editor to evaluate the prototype performance. Figure 6.23a show the board's simplified schematic. The PMU can be directly connected to an input DC voltage generator (V_{DC}), or a variable voltage source, such as a supercapacitor.

---

[3] https://www.autodesk.com/products/eagle/overview.

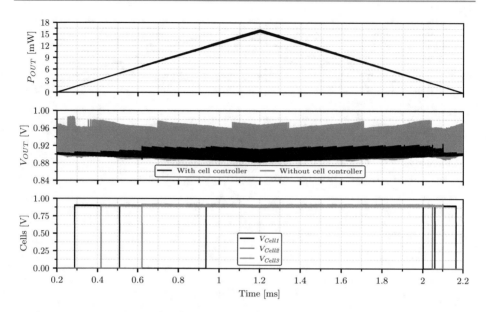

Fig. 6.18 Schematic simulation results: PMU transient response to an output current ramp signal from 0 A to 17.78 mA (16 mW) and back to 0 A, with $V_{IN} = 1.95$ V

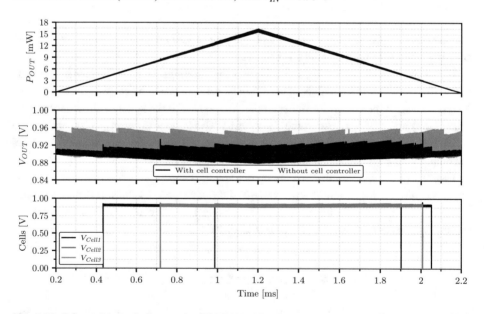

Fig. 6.19 Schematic simulation results: PMU transient response to an output current ramp signal from 0 A to 17.78 mA (16 mW) and back to 0 A, with $V_{IN} = 1.8$ V

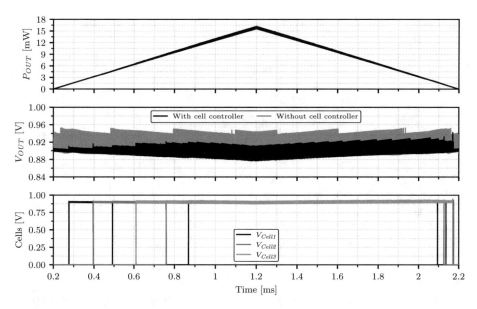

Fig. 6.20 Schematic simulation results: PMU transient response to an output current ramp signal from 0 A to 17.78 mA (16 mW) and back to 0 A, with $V_{IN} = 1.55$ V

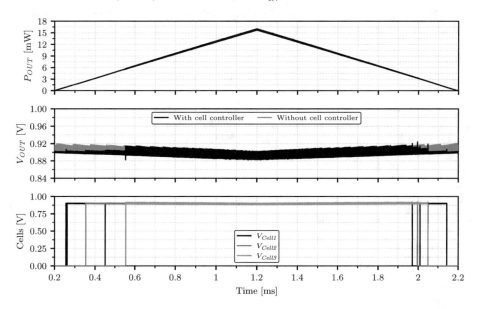

Fig. 6.21 Schematic simulation results: PMU transient response to an output current ramp signal from 0 A to 17.78 mA (16 mW) and back to 0 A, with $V_{IN} = 1.1$ V

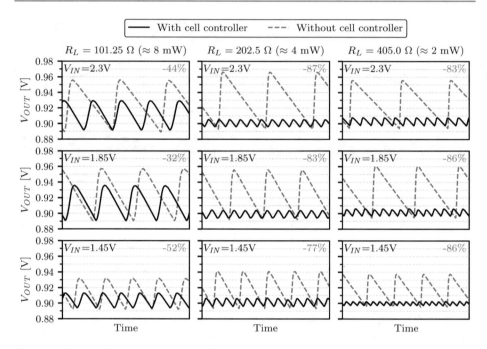

Fig. 6.22 Schematic simulation results: PMU output voltage ripple with and without the cell controller for different V_{IN} and P_{OUT} values

Table 6.1 Simulation results summary

Tech (nm)	130
CR #	3
CR	1/2, 2/3, 1/1
C_{FLY} (Type)	MOS
C_{OUT} (Type)	MOS
Phase interleaved	32
V_{IN} (V)	1.1–2.3
V_{IN} range (V)	1.2
V_{OUT} (V)	0.9
V_{OUT} range (V)	–
$V_{Ripple_{AVG}}$ (max) (mV)	47.7
Area (mm^2)	5.12
η_{max} (%)	85.1
P_{OUT} at η_{max} (mW)	16
$P_{density}$ at η_{max} (mW/mm^2)	3.125
$P_{density_{max}}$ (mW/mm^2)	3.125

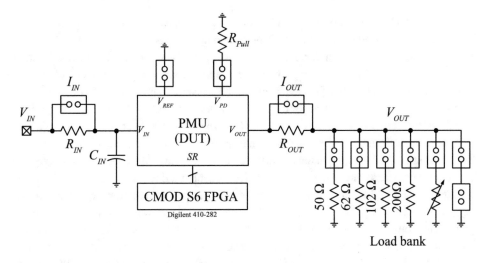

(a) Simplified Schematic.

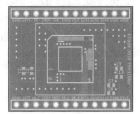

(b) Breakout board layout.

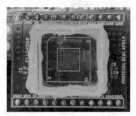

(c) Breakout board photography.

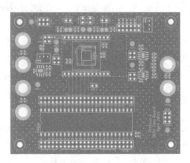

(d) Mainboard layout.

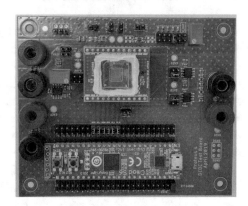

(e) Mainboard photography.

Fig. 6.23 PCB for the prototype evaluation

As for the PMU output, different load resistors, to test different output powers, are made available through headers. The PMU's input and output currents are measured by sensing the voltage drop across a $50\,\text{m}\Omega$ resistors, R_{IN}/R_{OUT}, that is in series with the input/output paths. This, along with the sensing of the input and output voltage, allows to obtain the input and output power, and thus determining the converter's efficiency. The power down pad V_{PD} has a pull down resistor for the PMU to work. An FPGA CMOD S6, by Digilent,[4] was used to communicate with the Set-Reset (SR) to change the reference voltage and also to acquire the cells and CR signals. Additional circuits such as voltage regulators of 0.9 and 3.3 V were also implemented in the board to power the pad ring. Moreover, extra decoupling capacitors footprints were placed in the board to solder them if required. Figure 6.23d shows the layout image of the 2 layers PCB, where the top layer was used for routing and the bottom layer for the ground plane. The unused space on the top layer was covered with a ground plane. The power traces were made wide enough and as short as possible, to minimize the parasitic resistances and inductances. Instead of attaching the die directly to the PCBs, a break-out-board (Fig. 6.23b) was used to attach it, which is then connected to the PCBs main board through pin heads. This facilitates the test of multiple samples without having to use one PCBs main board for each die. Measurements were performed with the PMU signals applied directly on the breakout board and besides a smaller reduction on the output voltage ripple, no other difference was noticed when compared with the measurements on the main board, with the breakout board attached. Figure 6.23c and e show the boards photography with the soldered components and the die. The die was glued to the PCB and its PADs were wire-bonded to the PCB. The die has a glass glued to the PCB for protection.

6.2.6 Prototype Measurements

During the layout process, an error occurred and the PMU's input/output pads were not disconnected from the ESD protection of the pad ring. Figure 6.24 shows the simplified diagram of the analogue I/O cells ESD protection, where V_{DDA} is the supply voltage for other circuits in the die. This pad should be connected to 0.9 V, which means that when the PMU $V_{IN} > V_{DDA}$ current will flow from V_{IN} to V_{DDA} through the ESD diodes, and since V_{DDA} is the supply voltage of other circuits, they will start to draw current and further increase the current consumption. Furthermore, the PMU's V_{OUT} may also rise over V_{DDA} due to the output voltage ripple.

To minimize the current leak from the ESDs diodes, a measurement was made with V_{DDA} connected to the same supply voltage as V_{IN}, i.e. $V_D = V_{DDA} - V_{IN} = 0$ V, therefore, ideally no current should flow across the ESDs diode. Figure 6.25 shows the PMU's input current into the V_{IN} pad, with no load connected to the PMU's output, and for V_{DDA} floating and connected to the V_{IN}. When V_{DDA} is left floating, the input current value increases as V_{IN} increases, reaching a total of 1.63 mA and an average voltage of 0.8 mA. When

---

[4] https://store.digilentinc.com/.

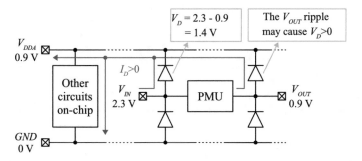

Fig. 6.24 Simplified diagram of the analogue I/O cells ESD protection

Fig. 6.25 Measurement
results: I_{IN} as a function of
V_{IN}, with V_{DDA} floating or
connected to the same supply
voltage as V_{IN}

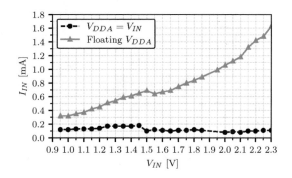

V_{DDA} is connected to the same supply voltage as V_{IN}, the average input current drops to
0.12 mA. This confirms that current is leaking through the ESDs diodes. Even though all
the measurements shown from now on were made with V_{DDA} connected to the same supply
voltage as V_{IN}, there is no absolute certainty that there is no leaking current through the
ESDs diodes. Meaning that it will not be possible to have an accurate measurement of the
PMU's efficiency. The region between 2.0 and 1.8 V has no points in the graph because
oscillations between the 1/2 and the 2/3 CR were observed. Hence, neither the input current
nor the voltage ripple was measured.

Figure 6.26 shows the PMU's prototype measurement results of the efficiency for different
V_{IN} values, and the comparison with the simulated ones. In the overall, 1/1 CR is the one
that has the efficiency values close to the simulation values. The 2/3 CR is far from the
simulation values for low R_L values, however, it seems to have better results in higher R_L
values, e.g. $R_L = 200\ \Omega$. The 1/2 CR has the worst results, where the PMU is not able
to produce a stable output voltage below an input voltage of 2.1 V, and thus the efficiency
drops rapidly. Still, for input voltage values above 2.1 V, the efficiency values are not far
from the simulated ones. Table 6.2 depicts the average, maximum, and minimum values of
the simulated and measured efficiency values. The results show that at low power levels,
the measured efficiency gets closer to the simulated one. Furthermore, the PMU is able to

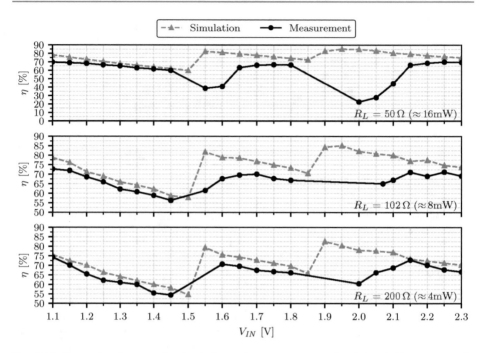

Fig. 6.26 Comparison between the simulation and measurement results of the PMU prototype efficiency

Table 6.2 Comparison of the measurement and simulation results of the PMU's efficiency

	R_L (Ω)	η_{Sim} (%)	η_{Meas} (%)	$\eta_{Meas} - \eta_{Sim}$ (%)
Avg.	50 ($\approx$ 16 mW)	75.32	58.74	−16.58
	102 ($\approx$ 8 mW)	74.11	67.35	−6.76
	200 ($\approx$ 4 mW)	70.99	66.30	−4.69
Max.	50 ($\approx$ 16 mW)	85.09	69.93	−15.16
	102 ($\approx$ 8 mW)	85.11	72.70	−12.41
	200 ($\approx$ 4 mW)	82.55	74.30	−8.25
Min.	50 ($\approx$ 16 mW)	59.90	22.62	−37.28
	102 ($\approx$ 8 mW)	57.81	56.30	−1.51
	200 ($\approx$ 4 mW)	54.60	54.40	−0.20

work until its CR voltage limit in the 1/1 and 2/3 CR. Only in the 1/2 CR the output voltage decreases rapidly as approaching the voltage limit.

Figure 6.27 shows the values of the peak to peak measurement on the oscilloscope of the PMU's output voltage for four different load values 50, 68, 102, 200, and 405 Ω, which at 0.9 V gives an approximate P_{OUT} value of 16, 12, 8, 4, and 2 mW. The graph shows that

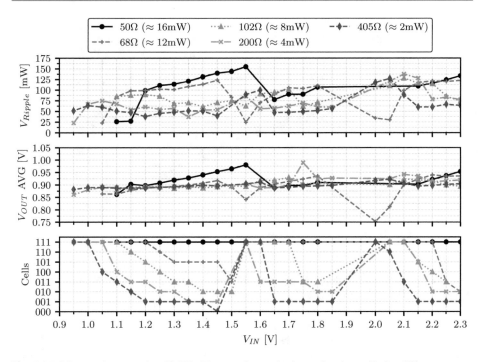

Fig. 6.27 Measurement results: PMU's V_{OUT} voltage, ripple, and active cells for different output loads as a function of V_{IN}

the output voltage ripple is higher than the simulated values. Reaching values of 150 mV when $R_L = 50\ \Omega$. Nonetheless, it can be seen that when the PMU is able to reduce the number of cells at lower power levels, and thus the voltage ripple is decreased. For example, for $R_L = 405\ \Omega$, when V_{IN} is higher than 2.15 V, the voltage ripple is around 60 mV with the cell code of 001. At 2.1, 2.05, and 2.0 V the number of enabled cells increases from 001 to 011, 110, and finally to 111. The voltage ripple is now at 120 mV, the double of the initial value. The mean value of the ripple is 106.39 mV for $R_L = 50\ \Omega$, 90.61 V for the $R_L = 68\ \Omega$, 85.79 mV for $R_L = 102\ \Omega$, 69.91 for $R_L = 200\ \Omega$, and 62.8 mV for the $R_L = 405\ \Omega$. These numbers show a voltage ripple reduction at lower power levels. The graph also shows that the converter is able to keep the output voltage regulated at 0.9 V.

Figure 6.28 shows the PMU's V_{OUT} calibration with $V_{IN} = 2.3$ V and without any load resistor. The V_{REF} from the voltage reference generator was swept, using a button on the FPGA, that increments a resistor in the bandgap ladder, until all are enabled. If pressed again, all the resistors are disabled, and then starts enabling again. The figure shows that the output voltage can vary approximately 200 mV, from 1 to 0.8 V.

Figure 6.29 shows the PMU's V_{OUT} when starting up. The figures show that the PMU can successfully start up, however, when the V_{IN} ramp is slow, there are some oscillations in the

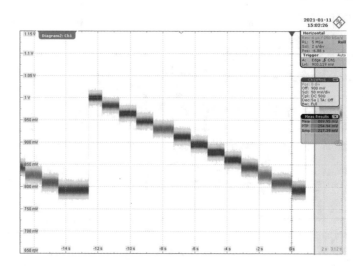

Fig. 6.28 Measurement results: PMU's voltage reference generator calibration reflected on V_{OUT} at $V_{IN} = 2.3$ V with no load and with V_{DDA} connected to V_{IN}

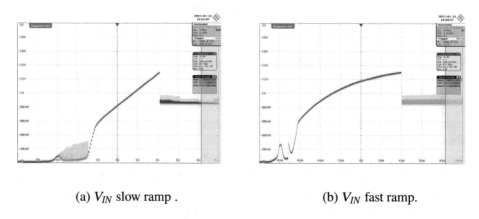

(a) V_{IN} slow ramp .   (b) V_{IN} fast ramp.

Fig. 6.29 Measurement results: PMU start-up V_{OUT} waveform

PMU's V_{OUT} whilst V_{IN} is lower than 200 mV, however, this does not prevent a successful start-up.

Figure 6.30 shows the PMU's load step response from a no load condition to 50 Ω and back to no load. Whilst the voltage droop is not far from the simulated one, the response time is slower, it takes approximately 400 ns to recover from the voltage droop. As for the step from 50 Ω to no load, it took 173 μs to recover from the step, which again is larger than the simulation value.

Figures 6.31 and 6.32 shows the measurement results of the PMU's V_{OUT}, for different V_{IN} values, with no load resistor connected and with a 50 Ω load resistor, respectively. The

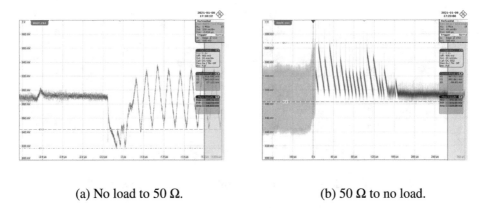

(a) No load to 50 Ω.   (b) 50 Ω to no load.

Fig. 6.30 Measurement results: load step response (no load to 50 Ω and back to no load) at $V_{IN} =$ 1.2 V

V_{IN} values were chosen so that they were in the maximum and minimum value of each CR. Finally, Table 6.3 shows the simulated and measurements results comparison.

The measured decrease in the converter's output voltage, and consequently the decrease in the converter's efficiency, when approaching the CR voltage limits at maximum output power, might be explained by the wires resistance and associated parasitic capacitance. Due to the large size of the layout of the circuit (3.2×1.6 mm) the wires used for V_{DD}, V_{SS} and the clock signals are necessarily long, which can result in large parasitic capacitance and series resistances.

The layout of the PMU has the control circuits on the left side and the Switched Capacitor (SC) converter cells on the right side, as shown in the simplified view of the layout of the circuit, depicted in Fig. 6.33. In the control circuits, the phase generator generates the 32 clock signals required to operate the 32-time interleaved SC converter cells. These 32 clock signals are transformed into 32 pairs of non-overlapping clock phases (ϕ_1 and ϕ_2) by the Non-overlap Clock Generator (NCG) block and then sent to the SC converter cells using a bus in the middle of the layout (shown in black in Fig. 6.33). The parasitic capacitance and resistance of the wires in this bus can change the non-overlapping time of the clock phases, creating time instants where two switches can short V_{DD} to V_{SS}. Due to time constraints and limited computation resources, it was not possible to simulate the extraction of the PMU circuit before tape-out. All the schematic simulations showed the circuit operating perfectly.

In order to investigate this issue, a simulation with an R-C-CC extraction of NCG, which includes the long clock buses represented in black lines between cell 4 and cells 1, 2, and 3 in Fig. 6.33; and schematic view of the phase generator was performed. Figure 6.34 shows the simulation results of the $\phi_{1<0>}$ and $\phi_{2<0>}$, which were observed at the right end of the clock lines. It can be seen that the pulses suffer some level of degradation and delay from the long clock lines, whereas in the schematic view this effect is not observed.

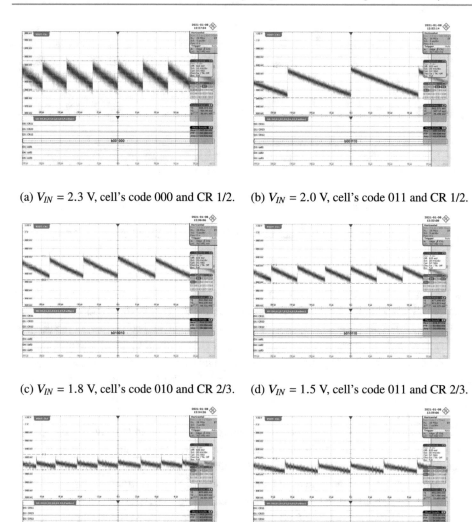

(a) V_{IN} = 2.3 V, cell's code 000 and CR 1/2.   (b) V_{IN} = 2.0 V, cell's code 011 and CR 1/2.

(c) V_{IN} = 1.8 V, cell's code 010 and CR 2/3.   (d) V_{IN} = 1.5 V, cell's code 011 and CR 2/3.

(e) V_{IN} = 1.45 V, cell's code 000 and CR 1/1.   (f) V_{IN} = 1.1 V, cell's code 010 and CR 1/1.

Fig. 6.31 Measurement results: PMU's V_{OUT} waveforms with no load connected

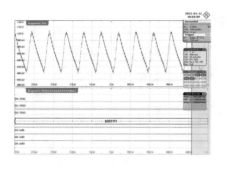
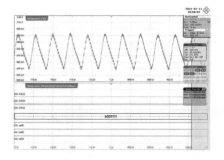

(a) V_{IN} = 2.3 V, cell's code 111 and CR 1/2.   (b) V_{IN} = 2.15 V, cell's code 111 and CR 1/2.

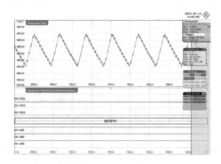
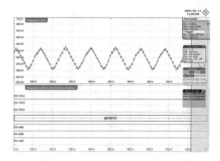

(c) V_{IN} = 1.8 V, cell's code 111 and CR 2/3.   (d) V_{IN} = 1.65 V, cell's code 111 and CR 2/3.

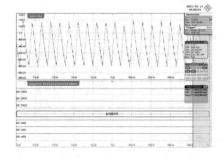
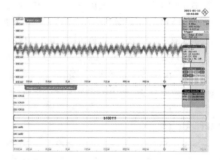

(e) V_{IN} = 1.5 V, cell's code 111 and CR 1/1.   (f) V_{IN} = 1.1 V, cell's code 111 and CR 1/1.

Fig. 6.32 Measurement results: PMU's V_{OUT} waveforms for $R_{LOAD} = 50\ \Omega$

Table 6.3 PMU's measurement and simulation results summary

Prototype	Simulation	Measurements
V_{IN} (V)	1.1–2.3	1.1–1.85, 2.0–2.3
V_{IN} range (V)	1.2	1.05
V_{OUT} (V)	0.9	0.9
V_{OUT} range (V)	–	–
$V_{Ripple_{AVG}}$ (max) (mV)	47.7	106.39
Area (mm^2)	5.12	5.12
η_{max} (%)	85.1	74.3
P_{OUT} at η_{max} (mW)	16	16
$P_{density}$ at η_{max} (mW/mm^2)	3.125	0.781
$P_{density_{max}}$ (mW/mm^2)	3.125	3.125

The same simulation was performed with the whole PMU with the schematic view, and the NCG and clock lines with R-C-CC extraction. Figure 6.35 shows the PMU's input current waveform, where spikes of 10 mA can be observed, demonstrating that there are two complementary clock phases active at the same time causing a short between V_{DD} and V_{SS}. These do not exist when the clock lines are in the schematic view. In this simulation, the current spikes were not enough to decrease the converter's output voltage. However, this can be because the schematic view of the cells does not take into account the resistances of the wires connecting the output of each cell to V_{DD}. Moreover, the NCG circuit is powered by V_{DD} and the lines connecting this circuit to V_{DD} can be closer to some SC cells than others, this can cause a drop in the V_{DD} voltage of the NCG, depending on the active SC cells. A drop in V_{DD} will necessarily affect the delay time between the non-overlapping phases, resulting in more shorts being created by the switches. This in turn causes more voltage drops in V_{DD}, which can ultimately cause the collapse of the PMU as observed in the measurements.

A last simulation was performed, where the resistance from the physical distance between the NCG and the converter cells was added. The NCG's V_{DD} and V_{SS} path resistance was added considering that both the metal traces have approximately 5 Ω each, from the supply frame to the circuit. Also, the resistance between the beginning of the converter's cells 3 and 4, until the middle point of each cell was added to the schematic, with the values shown in Fig. 6.36. Nevertheless, this value is exaggerated due to the matrix lines inside the converters.

The simulation results (Fig. 6.37) show that the V_{OUT} started to decrease after V_{IN} reaches 2 V. Also input current spikes were observed, which might be related with the overlap between two adjacent phases. The output voltage ripple has also increased, which can explain why the measured output voltage ripple is much larger than the simulation value.

To decrease the supply path physical resistance, and to decrease the clock buses length, the control circuits could have been placed in the middle of the cells. Also, a more careful

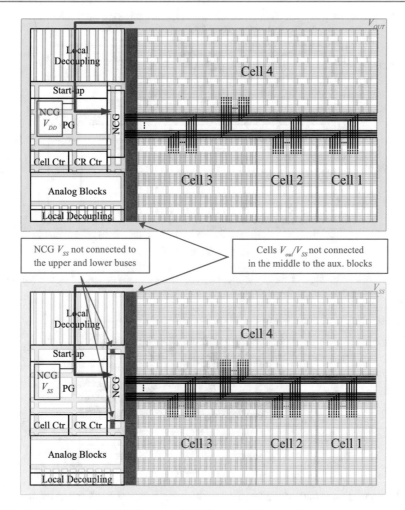

Fig. 6.33 Simplified layout block diagram of the V_{OUT} and V_{SS} metal traces

design of the supply lines should have been made, by using wider traces and with a metal stack, to further reduce the metal resistance. Another solution could be the use of a distributed clock generation scheme [1], where each cell receives the clock signal from the previous cell and pass it to the next cell. Thus, significantly reducing clock lines length.

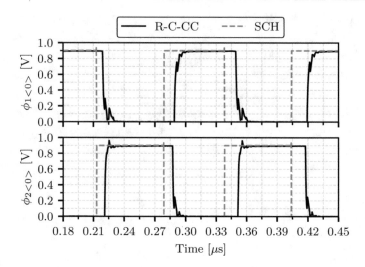

Fig. 6.34 Result simulation of the $\phi_{1<0>}$ and $\phi_{2<0>}$ clock signals generated by the NCG with the schematic view and extracted R-C-CC (including the long clock traces)

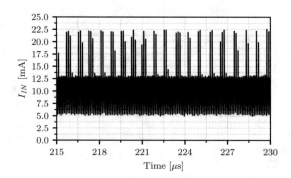

Fig. 6.35 Simulation results of the PMU schematic view with the NCG and clock lines R-C-CC extraction

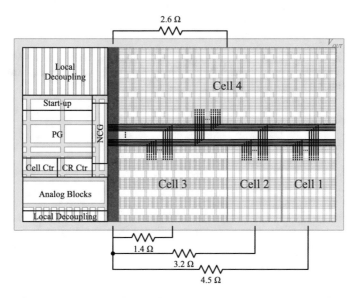

Fig. 6.36 Simplified layout block diagram of the V_{OUT} metal traces with the cells resistance from each cell

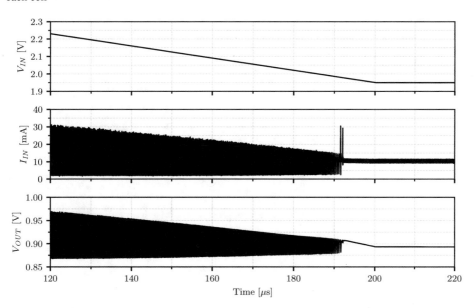

Fig. 6.37 Simulation results of the PMU schematic view with the NCG and clock lines R-C-CC extraction and the cells and NCG supply resistances

Reference

1. Lu Y, Jiang J, Ki W (2018) Design considerations of distributed and centralized switched-capacitor converters for power supply on-chip. IEEE J Emerg Sel Top Power Electron. https://doi.org/10. 1109/JESTPE.2017.2747094

Conclusions and Future Work

7

7.1 Conclusions

This book describes the design challenges of implementing a fully integrated Power Management Unit (PMU). It provides a step-by-step analysis for designing a multi-ratio and multi-cell Switched Capacitor (SC) DC-DC converter for, whilst not limited, a variable input voltage range. It also describes in detail each auxiliary circuit used for the proper behaviour of the SC converter. A prototypes in bulk 130 nm Complementary Metal-Oxide-Semiconductor (CMOS) technology was fabricated, and measurement results were depicted of the PMU's behaviour in different conditions. The following paragraphs will compare the prototype performance with the multi-ratio SC State-of-the-Art (SoA).

Table 7.1 shows the overall summary of the fully integrated PMU's described in this book. The prototype was designed for a maximum output power of 16 mW, for an input voltage range from 1.1 to 1.2 V, and an output voltage of 0.9 V. The SC converter is composed of three different Conversion Ratio (CR)s: 1/2, 2/3, and 1/1. The capacitors were implemented using MOS capacitors, and no external output capacitor was required. This was achieved by dividing the converter into 32 smaller cells that are clocked using a time interleaved scheme. Furthermore, capacitance modulation was employed, by dividing the interleaved converter into $1 + 3$ binary-weighted cells, that are enabled, or disabled, according to the output power demand and input voltage, both sensed through the clock phase signals. The total circuit active area is 5.12 mm^2. The auxiliary circuits, used for the controlling the converter's operation were: a Phase Generator (PG), a CR controller, a cell controller, the switched drivers, a voltage reference generator, and a start-up circuit. With these auxiliary circuits, the PMU is completely fully integrated.

Table 7.2 shows the comparison between the developed PMU's prototype with similar fully integrated multi-ratio SC converters, implemented in bulk CMOS technology, using the

© The Author(s), under exclusive license to Springer Nature Switzerland AG 2022
R. Madeira et al., *Fully Integrated Switched-Capacitor PMU for IoT Nodes*, Synthesis
Lectures on Engineering, Science, and Technology,
https://doi.org/10.1007/978-3-031-14701-2_7

Table 7.1 Overall summary of the PMU's prototypes

Prototype	PMU 16 mW	
Technology (nm)	130	
CR #	3	
CR	1/2, 2/3, 1/1	
Switch #	9	
C_{FLY} (Type)	MOS	
C_{OUT} (Type)	MOS	
C_{OUT} (nF)	4.7	
Area (mm^2)	5.12	
Phase interleaved	32	
Multi-cell	1+3 binary-weighted	
Results	Simulated	Measured
V_{in} (V)	2.3–1.1	1.1–1.85, 2.0–2.3
V_{in} range (V)	1.2	1.05
V_{out} (V)	0.9	0.9
V_{out} range (V)	–	–
V_{Ripple} (mV)	47.7*	106.39*
η_{max} (%)	85.1	74.3
P_{OUT} at η_{max} (mW)	16	16
$P_{density}$ at η_{max} (mW/mm^2)	3.125	0.781
$P_{density_{max}}$ (mW/mm^2)	3.125	3.125

*Maximum average value

Series-Parallel (SP) topology, and having step-down CRs. Figure 7.1 shows the prototype positioning in the fully integrated multi-ratio SC converters' SoA considering only the step-down converters implemented in bulk CMOS technology.

7.2   Future Work

During the design of the fully integrated PMU in CMOS technology, the following interesting considerations for futures work can be explored:

- The use of distributive clock generation instead of centralized [5]. Typically, the phase generator produces the clock phases required for the SC converter and then they are distributed to the converter cells. However, the clock generation could be distributed along the converter cells, where the inverter that generates the clock phase is implemented in each converter cell and the adjacent converter cell adds the delay to the clock phase.

Table 7.2 PMU's prototypes comparison with similar SoA converters

Prototype	PMU 16 mW	[1]'13	[2]'15	[3]'17	[4]'18
Tech (nm)	130	130	65	65	180
CR #	3	2	3	3	4
CR	1/2, 2/3, 1/1	Down: 1/2, 1/3	Down 1/2, 2/3, 1/1 Up: 3/2, 2/1, 3/1	Down: 1/2, 2/3, 3/4	Down: 2/3, 3/5, 1/2, 2/5
Switch #	9	8	N.A.	12	14
C_{FLY} (Type)	MOS	MIM	N.R.	MOS, MOM, MIM	MIM
C_{OUT} (Type)	MOS	MOS, MIM	N.R.	MOS, MOM, MIM	MOS
C_{OUT} (nF)	4.7	5	N.R.	2	5
Area (mm^2)	5.12	0.97	0.48	0.84	1.06
Phase interleaved	32	Yes (2)	No	Yes (123)	No
V_{in} (V)	1.1–1.85, 2.0–2.3	1.2	0.5–3.3	1.2–2.2	2–3.6
V_{in} range (V)	1.05	–	2.5	0.6	1.6
V_{out} (V)	0.9	0.3–0.55	1	0.6 - 1.2	1.2
V_{out} range (V)	–	0.25	–	0.6	–
V_{Ripple} (mV)	106.39[1]	50	N.R.	30	90
η_{max} (%)	74.3	70	Down: 69* Up: 70.4	80	84.2
P_{OUT} at η_{max} (mW)	16	24.5	0.0034	56	2.4
$P_{density}$ at η_{max} (mW/mm^2)	0.781	3.94*	0.007*	66.6	2.26
$P_{density_{max}}$ (mW/mm^2)	3.125	24.5	0.007*	180	2.40

*Estimated from the corresponding literature
[a]Maximum average value

Thus, forming a ring oscillator. This allows the use of other layout configurations for the converter's cells, instead of the rectangular shape due to the symmetrical layout for the centralized scheme. Also, the distributive paths are smaller resulting in lower power consumption on the phase generation [6];

- The use of multiple output voltages from a single converter. The System-on-Chip (SoC) typically requires different voltage domains due to the different circuit types in it. Since the fully integrated SC converter's area is large, especially when compared with the SoC area, it is important to maximize the number of circuits being powered by the same converter to reduce the area. Instead of having a single converter for each voltage domain [7];
- The development of an optimization tool that automatically generates a multi-ratio converter out of a set of specifications. Combining multiple topologies in a single multi-ratio SC converter can be a challenging task. It would be interesting to develop a tool that automatically generates the multi-ratio converter, for example using a genetic algorith-

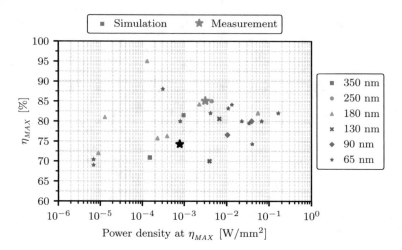

Fig. 7.1 Prototype position in the SoA of fully integrated multi-ratio SC converters considering bulk CMOS technologies: maximum efficiency versus respective power density

mic optimization to find the best solution. The tool would return the converter with all its components chosen and sized, from a set of specifications, such as the input and output voltage ranges, maximum output power, and so on. This would ease the design complexity for the designer.

References

1. Kudva SS, Harjani R (2013) Fully integrated capacitive DC-DC converter with all-digital ripple mitigation technique. IEEE J Solid-State Circuits. https://doi.org/10.1109/JSSC.2013.2259044
2. Hua X, Harjani R (2015) 3.5–0.5 V input, 1.0 V output multi-mode power transformer for a supercapacitor power source with a peak efficiency of 70.4%. In: 2015 IEEE custom integrated circuits conference (CICC), https://doi.org/10.1109/CICC.2015.7338390
3. Lu Y, Jiang J, Ki W (2017) A multiphase switched-capacitor DC-DC converter ring with fast transient response and small ripple. IEEE J Solid-State Circuits. https://doi.org/10.1109/JSSC.2016.2617315
4. Ianxi L, Hao C, Tianyuan H, Junchao M, Zhangming Z, Yintang Y (2018) A dual mode step-down switched-capacitor DC-DC converter with adaptive switch width modulation. Microelectron J. https://doi.org/10.1016/j.mejo.2018.06.003
5. Lu Y, Jiang J, Ki W (2018) Design considerations of distributed and centralized switched-capacitor converters for power supply on-chip. IEEE J Emerg Sel Top Power Electron. https://doi.org/10.1109/JESTPE.2017.2747094
6. Jiang J, Liu X, Mok Ki WH, PKT, Lu Y (2021) Circuit techniques for high efficiency fully-integrated switched-capacitor converters. IEEE Trans Circuits Syst II: Express Briefs. https://doi.org/10.1109/TCSII.2020.3046514

7. Jiang J, Lu Y, Ki W, U S, Martins RP (2017) A dual-symmetrical-output switched-capacitor converter with dynamic power cells and minimized cross regulation for application processors in 28 nm CMOS. In: IEEE international solid-state circuits conference (ISSCC). https://doi.org/10.1109/ISSCC.2017.7870402

Appendix

A

A.1 PMU 16 mW Design

Figure A.1 shows the simplified form of the Power Management Unit (PMU) multi-ratio Series-Parallel (SP) Switched Capacitor (SC) converter used to implement the Conversion Ratios (CRs) of 1/2, 2/3, and 1/1. The following sections will describe the converter design equations for each CR and how they were used to size its passive components.

A.1.1 Design Equations of the 1/2 CR

Figure A.2 shows the converter in the 1/2 CR configuration, for each phase. Assuming that V_{OUT} is kept at a constant voltage, the charge equations can be drawn:

$\phi_1 \rightarrow \phi_2$:

$$(V_{IN} - V_{OUT})(C_1 + C_2) + V_{IN}(\alpha_1 C_1 + \alpha_2 C_2) = V_{OUT}(C_1 + C_2 + \alpha_1 C_1 + \alpha_2 C_2) + \Delta q_o^{\phi_2} \quad \text{(A.1)}$$

$\phi_2 \rightarrow \phi_1$:

$$- V_{OUT}(C_1 + C_2) = (V_{OUT} - V_{IN})(C_1 + C_2) + V_{OUT}(\beta_1 C_1 + \beta_2 C_2) + \Delta q_o^{\phi_1} \quad \text{(A.2)}$$

$$V_{OUT}(C_1 + C_2 + \alpha_1 C_1 + \alpha_2 C_2) = (V_{IN} - V_{OUT})(C_1 + C_2) + V_{IN}(\alpha_1 C_1 \alpha_2 C_2) - \Delta q_i^{\phi_1} \quad \text{(A.3)}$$

where $\Delta q_o^{\phi_{1,2}}$ are the amount of charge absorbed by V_{OUT}, in the respective phase, and $\Delta q_i^{\phi_1}$ the amount of charge drawn by the circuit from V_{IN}, in this case only during ϕ_1.

© The Editor(s) (if applicable) and The Author(s), under exclusive license to Springer
Nature Switzerland AG 2022
R. Madeira et al., *Fully Integrated Switched-Capacitor PMU for IoT Nodes*, Synthesis
Lectures on Engineering, Science, and Technology,
https://doi.org/10.1007/978-3-031-14701-2

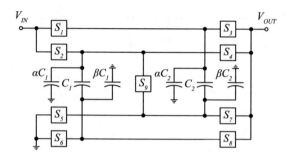

Fig. A.1 Simplified schematic of the PMU multi-ratio SC converter

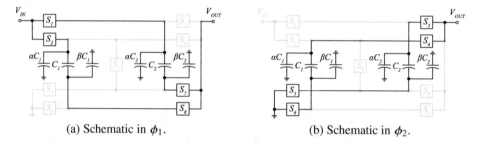

(a) Schematic in ϕ_1.   (b) Schematic in ϕ_2.

Fig. A.2 Simplified schematic of the converter in the 1/2 CR configuration

These equations can be solved in respect to $\Delta q_i^{\phi_1}$, $\Delta q_o^{\phi_1}$, and $\Delta q_o^{\phi_2}$, which results in

$$\Delta q_i^{\phi_1} = C_1 V_{IN} (1 + \alpha_1) - C_1 V_{OUT} (2 + \alpha_1) + C_2 V_{IN} (1 + \alpha_2) - C_2 V_{OUT} (2 + \alpha_2) \quad \text{(A.4)}$$

$$\Delta q_o^{\phi_1} = C_1 (V_{IN} - V_{OUT} (2 + \beta_1)) + C_2 (V_{IN} - V_{OUT} (2 + \beta_2)) \quad \text{(A.5)}$$

$$\Delta q_o^{\phi_2} = C_1 V_{IN} (1 + \alpha_1) - C_1 V_{OUT}(2 + \alpha_1) + C_2 V_{IN} (1 + \alpha_2) - C_2 V_{OUT} (2 + \alpha_2) \quad \text{(A.6)}$$

For simplicity, and since the flying capacitors are to be implemented by the same device, it is assumed that $C_1 = C_2 = C_{FLY}$, $\alpha_1 = \alpha_2 = \alpha$, and $\beta_1 = \beta_2 = \beta$. Hence, the input and output current, and power can then be determined by

$$I_{IN} = \Delta q_i^{\phi_1} F_{CLK} = 2 F_{CLK} C_{FLY} (V_{IN} (1 + \alpha) - V_{OUT} (2 + \alpha)) \quad \text{(A.7)}$$

$$I_{OUT} = (\Delta q_o^{\phi_1} + \Delta q_o^{\phi_2}) F_{CLK} = 2 F_{CLK} C_{FLY} (V_{IN} (2 + \alpha) - V_{OUT} (4 + \alpha + \beta)) \quad \text{(A.8)}$$

$$P_{IN} = V_{IN} I_{IN} = 2 C_{FLY} F_{CLK} V_{IN} (V_{IN} (1 + \alpha) - V_{OUT} (2 + \alpha)) \quad \text{(A.9)}$$

$$P_{OUT} = V_{OUT} I_{OUT} = 2 C_{FLY} F_{CLK} V_{OUT} (V_{IN} (2 + \alpha) - V_{OUT} (4 + \alpha + \beta)) \quad \text{(A.10)}$$

The converter efficiency (η) can be obtained by

$$\eta = \frac{P_{OUT}}{P_{IN}} = \frac{V_{OUT} (V_{IN} (2 + \alpha) - V_{OUT} (4 + \alpha + \beta))}{V_{IN} (V_{IN} (1 + \alpha) - V_{OUT} (2 + \alpha))} \quad \text{(A.11)}$$

The ideal CR and output impedance (R_{OUT}) of the converter can be determined by

$$CR\bigg|_{\alpha,\beta=0} = \frac{I_{IN}}{I_{OUT}} = \frac{1}{2} \tag{A.12}$$

$$R_{OUT}\bigg|_{\alpha,\beta=0} = \frac{1}{8\,C_{FLY}\,F_{CLK}} \tag{A.13}$$

Both the converter's output voltage (V_{OUT}) and clock frequency (F_{CLK}) can be determined by

$$V_{OUT} = I_{OUT}\,R_L \Rightarrow V_{OUT} = \frac{2\,C_{FLY}\,F_{CLK}\,R_L\,V_{IN}\,(2+\alpha)}{1 + 2\,C_{FLY}\,F_{CLK}\,R_L\,(4+\alpha+\beta)} \tag{A.14}$$

$$F_{CLK} = \frac{V_{OUT}}{2\,C_{FLY}\,R_L\,(V_{IN}\,(2+\alpha) - V_{OUT}\,(4+\alpha+\beta))} \tag{A.15}$$

$$F_{CLK} = \frac{P_{OUT}}{2\,C_{FLY}\,V_{OUT}\,(V_{IN}\,(2+\alpha) - V_{OUT}\,(4+\alpha+\beta))} \tag{A.16}$$

where R_L is the load resistor.

Figure A.3 shows the simplified schematic of the converter in the 1/2 CR configuration, where the switches were replaced by resistors, which have a low R_{ON} value when ON, or a high R_{OFF} value when OFF. In this CR, both ϕ_1 (Fig. A.5a) and ϕ_2 (Fig. A.5b) have the same time constant, given by

$$\tau = R_{ON_{tot}}C_{FLY} = \frac{1}{2\,N\,F_{CLK}} \tag{A.17}$$

where C_{FLY} is equal to C_1 in one branch, and C_2 in the other branch, and $R_{ON_{tot}}$ is the total resistance value in each branch per phase, which is equal to $R_{ON_{tot}} = R_{ON\,S1} + R_{ON\,S7}$ (or $R_{ON\,S2} + R_{ON\,S8}$) in ϕ_1, and $R_{ON_{tot}} = R_{ON\,S3} + R_{ON\,S5}$ (or $R_{ON\,S4} + R_{ON\,S6}$) in ϕ_2.

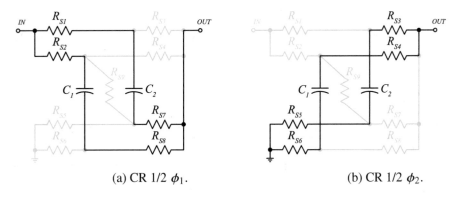

(a) CR 1/2 ϕ_1.   (b) CR 1/2 ϕ_2.

Fig. A.3 Simplified schematic of the converter in the 1/2 CR configuration with the switches replaced by resistors

Assuming that all the resistors have the same value, R_{ON}, then $R_{ON} = R_{ON_{tot}}/2$ for each branch, in both phases. Hence, considering a settling time of four time constants ($N = 4$), R_{ON} is given by

$$R_{ON_{tot}} = \frac{1}{2\,N\,C_{FLY}\,F_{CLK}} \Rightarrow R_{ON} = \frac{1}{16\,C_{FLY}\,F_{CLK}} \tag{A.18}$$

Since the coefficients k_R and k_C of the PMOS are much larger than the NMOS, the paths which have both transistors in series were sized so that the area was distributed between both. That is, the NMOS was oversized, so that the PMOS would be reduced. The following equations show how R_{ON} was sized for each switch, in the 1/2 CR configuration.

$$F_{CLK12_{MAX}} = \frac{P_{OUT}}{C_{FLY}\,V_{OUT}\,(V_{IN_{Lim}}\,(2+\alpha) - V_{OUT}(4+\alpha+\beta))} \tag{A.19}$$

$$\phi_1 = \begin{cases} R_{S2} = R_{S8} \\ R_{S2} + R_{S8} = 1/(8\,C_1\,F_{CLK12_{MAX}}) \\ R_{S1} = R_{S7} \\ R_{S1} + R_{S7} = 1/(8\,C_2\,F_{CLK12_{MAX}}) \end{cases} \qquad \phi_2 = \begin{cases} R_{S4} = \frac{k_{R12N}}{k_{R12P}}\,R_{S6} \\ R_{S4} + R_{S6} = 1/(8\,C_1\,F_{CLK12_{MAX}}) \\ R_{S3} = \frac{k_{R12N}}{k_{R12P}}\,R_{S5} \\ R_{S3} + R_{S5} = 1/(8\,C_2\,F_{CLK12_{MAX}}) \end{cases}$$

A.1.2   Design Equations of the 2/3 CR

Figure A.4 shows the converter in the 2/3 CR configuration, for each phase. Assuming that V_{OUT} is kept at a constant voltage, the charge equations can be drawn: $\phi_2 \to \phi_1$:

$$(V_{OUT} - V_{IN})\,C_2 + (V_{IN} - V_{OUT})\,C_1 + V_{IN}\,\alpha_1\,C_1 + V_{OUT}\,\beta_2\,C_2 =$$
$$= (V_x - V_{OUT})\,C_2 + V_x\,(C_1 + \beta_2\,C_2 + \alpha_1\,C_1) \tag{A.20}$$

$$(V_{IN} - V_{OUT})\,C_2 + V_{IN}\,\alpha_2\,C_2 = (V_{OUT} - V_x)\,C_2 + V_{OUT}\,\alpha_2\,C_2 + \Delta q_o^{\phi_2} \tag{A.21}$$

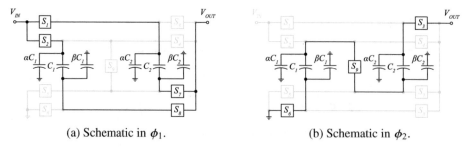

(a) Schematic in ϕ_1.   (b) Schematic in ϕ_2.

Fig. A.4 Simplified schematic of the converter in the 2/3 CR configuration

$$(V_{OUT} - V_x)\, C_2 + V_x\, C_1 + V_x\, \alpha_1\, C_1 + V_{OUT}\, \alpha_2\, C_2 =$$
$$= (V_{IN} - V_{OUT})\, (C_1 + C_2) + V_{IN}\, (\alpha_1\, C_1 + \alpha_2\, C_2) - \Delta q_i^{\phi_1} \quad \text{(A.22)}$$

$\phi_1 \rightarrow \phi_2$:

$$(V_x - V_{OUT})\, C_2 - V_x\, C_1 + V_x\, \beta_2\, C_2 =$$
$$(V_{OUT} - V_{in})\, (C_1 + C_2) + V_{OUT}\, (\beta_1\, C_1 + \beta_2\, C_2) + \Delta q_o^{\phi_1} \quad \text{(A.23)}$$

where $\Delta q_o^{\phi_{1,2}}$ are the amount of charge absorbed by V_{OUT}, in the respective phase, and $\Delta q_i^{\phi_1}$ the amount of charge drawn by the circuit from V_{IN}, in this case only during ϕ_1.

These equations can be solved in respect to $\Delta q_i^{\phi_1}$, $\Delta q_o^{\phi_1}$, and $\Delta q_o^{\phi_2}$, which for simplicity it is assumed that both the capacitors are equal, thus that $C_1 = C_2 = C_{FLY}$, $\alpha_1 = \alpha_2 = \alpha$, and $\beta_1 = \beta_2 = \beta$. This results in

$$\Delta q_i^{\phi_1} = \frac{C_{FLY}\, (V_{IN}\, (\alpha^2 + 2\alpha\, (\beta + 3) + 2\, (\beta + 2)) - V_{OUT}\, (\alpha^2 + 2\alpha\, (\beta + 3) + 3\, (\beta + 2)))}{2 + \alpha + \beta} \quad \text{(A.24)}$$

$$\Delta q_o^{\phi_1} = \frac{C_{FLY}\, (V_{IN}\, (\alpha\, (\beta + 2) + 2\, (\beta + 2)) - V_{OUT}\, (\alpha\, (2\beta + 3) + \beta\, (\beta + 6) + 6))}{2 + \alpha + \beta} \quad \text{(A.25)}$$

$$\Delta q_o^{\phi_2} = \frac{C_{FLY}\, (V_{IN}\, (\alpha^2 + \alpha\, (\beta + 4) + \beta + 2) - V_{OUT}\, (\alpha^2 + \alpha\, (\beta + 4) + \beta + 3))}{2 + \alpha + \beta} \quad \text{(A.26)}$$

The input and output current, and power can then be determined by

$$I_{IN} = \frac{C_{FLY}\, F_{CLK}\, (V_{IN}\, (\alpha^2 + 2\alpha\, (\beta + 3) + 2\, (\beta + 2)) - V_{OUT}\, (\alpha^2 + 2\alpha\, (\beta + 3) + 3\, (\beta + 2)))}{2 + \alpha + \beta} \quad \text{(A.27)}$$

$$I_{OUT} = \frac{C_{FLY}\, F_{CLK}\, (V_{IN}\, (\alpha^2 + 2\alpha\, (\beta + 3) + 3\, (\beta + 2)) - V_{OUT}\, (\alpha^2 + \alpha\, (3\beta + 7) + \beta\, (\beta + 7) + 9))}{2 + \alpha + \beta} \quad \text{(A.28)}$$

$$P_{IN} = V_{IN}\, I_{IN} =$$
$$= \frac{C_{FLY}\, F_{CLK}\, V_{IN}\, (V_{IN}\, (\alpha^2 + 2\alpha\, (\beta + 3) + 3\, (\beta + 2)) - V_{OUT}\, (\alpha^2 + \alpha\, (3\beta + 7) + \beta\, (\beta + 7) + 9))}{2 + \alpha + \beta} \quad \text{(A.29)}$$

$$P_{OUT} = V_{OUT}\, I_{OUT} =$$
$$= \frac{C_{FLY}\, F_{CLK}\, V_{OUT}\, (V_{IN}\, (\alpha^2 + 2\alpha\, (\beta + 3) + 2\, (\beta + 2)) - V_{OUT}\, (\alpha^2 + 2\alpha\, (\beta + 3) + 3\, (\beta + 2)))}{2 + \alpha + \beta} \quad \text{(A.30)}$$

The converter efficiency (η) can be obtained by

$$\eta = \frac{V_{OUT}\, (V_{IN}\, (\alpha^2 + 2\alpha\, (\beta + 3) + 3\, (\beta + 2)) - V_{OUT}\, (\alpha^2 + \alpha\, (3\beta + 7) + \beta\, (\beta + 7) + 9))}{V_{IN}\, (V_{IN}\, (\alpha^2 + 2\alpha(\beta + 3) + 2\, (\beta + 2)) - V_{OUT}\, (\alpha^2 + 2\alpha\, (\beta + 3) + 3\, (\beta + 2)))} \quad \text{(A.31)}$$

The ideal CR and output impedance (R_{OUT}) of the converter can be determined by

$$CR\Big|_{\alpha,\beta=0} = \frac{I_{IN}}{I_{OUT}} = \frac{2}{3} \tag{A.32}$$

$$R_{OUT}\Big|_{\alpha,\beta=0} = \frac{2}{9\,C_{FLY}\,F_{CLK}} \tag{A.33}$$

Both the output voltage and the clock frequency of the converter can be determined by

$$V_{OUT} = I_{OUT}\,R_L \Rightarrow V_{OUT} = \frac{C_{FLY}\,F_{CLK}\,R_L\,V_{IN}\,(\alpha^2 + 2\alpha\,(\beta+3) + 3\,(\beta+2))}{2 + \alpha + \beta + C_{FLY}\,F_{CLK}\,R_L\,(\alpha^2 + \alpha\,(3\beta+7) + \beta\,(\beta+7) + 9)} \tag{A.34}$$

$$F_{CLK} = \frac{V_{OUT}\,(2+\alpha+\beta)}{C_{FLY}\,R_L\,(V_{IN}\,(\alpha^2 + 2\alpha\,(\beta+3) + 3\,(\beta+2)) - V_{OUT}\,(\alpha^2 + \alpha\,(3\beta+7) + \beta\,(\beta+7) + 9))} \tag{A.35}$$

$$F_{CLK} = \frac{P_{OUT}\,(2+\alpha+\beta)}{C_{FLY}\,V_{OUT}\,(V_{IN}\,(\alpha^2 + 2\alpha\,(\beta+3) + 3\,(\beta+2)) - V_{OUT}\,(\alpha^2 + \alpha\,(3\beta+7) + \beta\,(\beta+7) + 9))} \tag{A.36}$$

where R_L is the load resistor.

Figure A.5 shows the simplified schematic of the converter in the 2/3 CR configuration, with the switches replaced by resistors, which have a low R_{ON} value when ON, or a high R_{OFF} value when OFF. In this CR both ϕ_1 (Fig. A.5a) and ϕ_2 (Fig. A.5b) have different time constants. They are given by

$$\phi_1 : \tau = R_{ON_{tot}}\,C_{FLY} = \frac{1}{2\,N\,F_{CLK}} \qquad \phi_2 : \tau = R_{ON_{tot}}\,C_{FLY} = \frac{1}{N\,F_{CLK}} \tag{A.37}$$

where $R_{ON_{tot}} = R_{ON\,S1} + R_{ON\,S7}$ (or $R_{ON\,S2} + R_{ON\,S8}$) in ϕ_1, and $R_{ON_{tot}} = R_{ON\,S3} + R_{ON\,S6} + R_{ON\,S9}$ in ϕ_2.

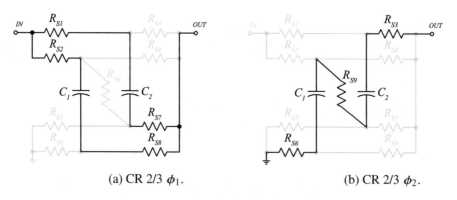

(a) CR 2/3 ϕ_1.   (b) CR 2/3 ϕ_2.

Fig. A.5 Simplified schematic of the converter in the 2/3 CR configuration with the switches replaced by resistors

Assuming that all the resistors have the same value, R_{ON}, then $R_{ON} = R_{ON_{tot}}/2$ in ϕ_1 for each branch, and $R_{ON} = R_{ON_{tot}}/3$ in ϕ_2. Hence, considering a settling time of four time constants ($N = 4$), R_{ON} is given by

$$\phi_1: \quad R_{ON_{tot}} = \frac{1}{8\,C_{FLY}\,F_{CLK}} \Rightarrow R_{ON} = \frac{1}{16\,C_{FLY}\,F_{CLK}} \tag{A.38}$$

$$\phi_2: \quad R_{ON_{tot}} = \frac{1}{4\,C_{FLY}\,F_{CLK}} \Rightarrow R_{ON} = \frac{1}{12\,C_{FLY}\,F_{CLK}} \tag{A.39}$$

Again, the NMOS and PMOS transistors were sized so that the area was distributed between both. The following equations show how R_{ON} was sized for each switch, in the 2/3 CR configuration.

$$F_{CLK23_{MAX}} = \tag{A.40}$$

$$= \frac{P_{OUT}\,(2 + \alpha + \beta)}{C_{FLY}\,V_{OUT}\left(V_{IN}\left(\alpha^2 + 2\,\alpha\,(\beta + 3) + 3\,(\beta + 2)\right) - V_{OUT}\left(\alpha^2 + \alpha\,(3\,\beta + 7) + \beta\,(\beta + 7) + 9\right)\right)} \tag{A.41}$$

$$\phi_1 = \begin{cases} R_{S2} = R_{S8} \\ R_{S2} + R_{S8} = 1/(8\,C_1\,F_{CLK23_{MAX}}) \\ R_{S1} = R_{S7} \\ R_{S1} + R_{S7} = 1/(8\,C_2\,F_{CLK23_{MAX}}) \end{cases} \qquad \phi_2 = \begin{cases} R_{S3} = \frac{k_{12N}}{W_x} \\ R_{S6} = \frac{k_{12P}}{W_x} \\ R_{S9} = \frac{k_{33N}}{W_x} \\ R_{S3} + R_{S6} + R_{S9} = 1/(4\,C_1\,F_{CLK23_{MAX}}) \end{cases}$$

A.1.3 Design Equations of the 1/1 CR

Figure A.6 shows the converter in the 1/1 CR configuration, for each phase. Assuming that V_{OUT} is kept at a constant voltage, the charge equations can be drawn:

$\phi_2 \to \phi_1$:

$$V_{IN}\,(C_1 + C_2 + \alpha_1\,C_1 + \alpha_2\,C_2) = V_{OUT}\,(C_1 + C_2 + \alpha_1\,C_1 + \alpha_2\,C_2) + \Delta q_o^{\phi_2} \tag{A.42}$$

$\phi_1 \to \phi_2$:

$$V_{OUT}\,(C_1 + C_2 + \alpha_1\,C_1 + \alpha_2\,C_2) = V_{IN}\,(C_1 + C_2 + \alpha_1\,C_1 + \alpha_2\,C_2) - \Delta q_i^{\phi_1} \tag{A.43}$$

where $\Delta q_o^{\phi_2}$ is the amount of charge absorbed by V_{OUT}, in the respective phase, and $\Delta q_i^{\phi_1}$ the amount of charge drawn by the circuit from V_{IN}, in this case only during ϕ_1.

These equations can be solved in respect to $\Delta q_i^{\phi_1}$ and $\Delta q_o^{\phi_2}$, which results in

$$\Delta q_i^{\phi_1} = (V_{IN} - V_{OUT})\,(C_1 + C_2 + \alpha_1\,C_1 + \alpha_2\,C_2) \tag{A.44}$$

$$\Delta q_o^{\phi_2} = (V_{IN} - V_{OUT})\,(C_1 + C_2 + \alpha_1\,C_1 + \alpha_2\,C_2) \tag{A.45}$$

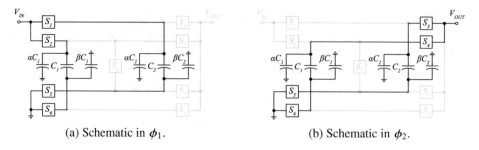

(a) Schematic in ϕ_1.   (b) Schematic in ϕ_2.

Fig. A.6 Simplified schematic of the converter in the 1/1 CR configuration

For simplicity, and since the flying capacitors are to be implemented by the same device, it is assumed that $C_1 = C_2 = C_{FLY}$, $\alpha_1 = \alpha_2 = \alpha$, and $\beta_1 = \beta_2 = \beta$. Hence, the input and output current, and power can then be determined by

$$I_{IN} = \Delta q_i^{\phi_1} F_{CLK} = 2 C_{FLY} F_{CLK} (V_{IN} - V_{OUT}) (1 + \alpha) \tag{A.46}$$

$$I_{OUT} = \Delta q_o^{\phi_2} F_{CLK} = 2 C_{FLY} F_{CLK} V_{IN} (V_{IN} - V_{OUT}) (1 + \alpha) \tag{A.47}$$

$$P_{IN} = V_{IN} I_{IN} = 2 C_{FLY} F_{CLK} (V_{IN} - V_{OUT}) (1 + \alpha) \tag{A.48}$$

$$P_{OUT} = V_{OUT} I_{OUT} = 2 C_{FLY} F_{CLK} V_{OUT} (V_{IN} - V_{OUT}) (1 + \alpha) \tag{A.49}$$

The converter efficiency (η) can be obtained by

$$\eta = \frac{P_{OUT}}{P_{IN}} = \frac{V_{OUT}}{V_{IN}} \tag{A.50}$$

The ideal CR and output impedance (R_{OUT}) of the converter can be determined by

$$CR \Big|_{\alpha,\beta=0} = \frac{I_{IN}}{I_{OUT}} = 1 \tag{A.51}$$

$$R_{OUT} \Big|_{\alpha,\beta=0} = \frac{1}{2 C_{FLY} F_{CLK}} \tag{A.52}$$

Both the output voltage and the clock frequency of the converter can be determined by

$$V_{OUT} = I_{OUT} R_L \Rightarrow V_{OUT} = \frac{2 C_{FLY} F_{CLK} R_L V_{IN} (1 + \alpha)}{1 + 2 C_{FLY} F_{CLK} R_L (1 + \alpha)} \tag{A.53}$$

$$F_{CLK} = \frac{V_{OUT}}{2 C_{FLY} R_L (V_{IN} - V_{OUT}) (1 + \alpha)} \tag{A.54}$$

$$F_{CLK} = \frac{P_{OUT}}{2 C_{FLY} (V_{IN} - V_{OUT}) V_{OUT} (1 + \alpha)} \tag{A.55}$$

where R_L is the load resistor.

Figure A.7 shows the simplified schematic of the converter in the 1/1 CR configuration, with the switches replaced by resistors, which have a low R_{ON} value when ON, or a high

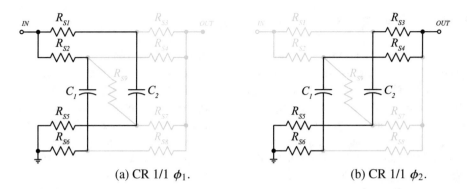

(a) CR 1/1 ϕ_1.   (b) CR 1/1 ϕ_2.

Fig. A.7 Simplified schematic of the converter in the 1/1 CR configuration with the switches replaced by resistors

R_{OFF} value when OFF. In this CR both ϕ_1 (Fig. A.7a) and ϕ_2 (Fig. A.7b) have the same number of elements in each path. Hence, the time constant is given by

$$\phi_{1,2} : \tau = R_{ON_{tot}} C_{FLY} = \frac{1}{2 N F_{CLK}} \tag{A.56}$$

where $R_{ON_{tot}} = R_{ON_{S1}} + R_{ON_{S5}}$ (or $R_{ON_{S2}} + R_{ON_{S6}}$) in ϕ_1, and $R_{ON_{tot}} = R_{ON_{S3}} + R_{ON_{S5}}$ (or $R_{ON_{S4}} + R_{ON_{S6}}$) in ϕ_2.

Assuming that all the resistors have the same value, R_{ON}, then $R_{ON} = R_{ON_{tot}}/2$ for each branch. Hence, considering a settling time of four time constants ($N = 4$), R_{ON} is given by

$$\phi_{1,2} : \quad R_{ON_{tot}} = \frac{1}{8 C_{FLY} F_{CLK}} \Rightarrow R_{ON} = \frac{1}{16 C_{FLY} F_{CLK}} \tag{A.57}$$

Again, the NMOS and PMOS transistors were sized so that the area was distributed between both. The following equation shows how the ON resistance was sized for the 1/1 CR.

$$F_{CLK11_{MAX}} = \frac{P_{OUT}}{2 C_{FLY} V_{OUT} (V_{IN} - V_{OUT}) (1 + \alpha)} \tag{A.58}$$

$$\phi_1 = \begin{cases} R_{S5} = \frac{k_{R12N}}{k_{R12P}} R_{S1} \\ R_{S1} + R_{S5} = 1/(8 C_2 F_{CLK11_{MAX}}) \\ R_{S6} = \frac{k_{R12N}}{k_{R12P}} R_{S2} \\ R_{S2} + R_{S6} = 1/(8 C_1 F_{CLK11_{MAX}}) \end{cases} \qquad \phi_2 = \begin{cases} R_{S3} + R_{S5_{\phi_1}} = 1/(8 C_2 F_{CLK11_{MAX}}) \\ R_{S4} + R_{S6_{\phi1}} = 1/(8 C_1 F_{CLK11_{MAX}}) \end{cases}$$

A.1.4 PMU Switches' Resistance Sizing

The converter is divided into 4 main cells that are further divided into 32 smaller interleaved cells. The total power of 16 mW is distributed between the four cells (cell 1, 2, 3, and 4) with the weights of 2, 2, 4, and 8 mW. Hence, an interleaved cell of cell 1 or 2, outputs 62.5 μW of power. The interleaved cells of cells 3 and 4 are made by placing two interleaved cells of 62.5 μW in parallel for cell 2, and four interleaved cells of 62.5 μW in parallel for cell 4. Therefore, to design the unitary interleaved cell, the input variables are $V_{OUT} = 0.9$ V, $P_{OUT} = 62.5$ μW, $C_{FLY} = 20$ pF ($\alpha = 3\%$ and $\beta \approx 0\%$), $V_{IN_{MIN\_CR12}} = 1.9$ V, $V_{IN_{MIN\_CR23}} = 1.5$ V, and $V_{IN_{MIN\_CR23}} = 1.1$ V, and finally $F_{CLK12_{MAX}} = 7.55$ MHz, $F_{CLK23_{MAX}} = 7.18$ MHz, and $F_{CLK11_{MAX}} = 8.43$ MHz. Table A.1

Table A.1 PMU 16 mW switches design in each CR configuration. Bold numbers are the worst case

Switch	R_{ON12} (Ω)	R_{ON23} (Ω)	R_{ON11} (Ω)	W_{12} (μm)	W_{23} (μm)	W_{11} (μm)	C_{GG12} (fF)	C_{GG23} (fF)	C_{GG11} (fF)
S_1	**414.00**	435.17	611.33	**6.54**	6.23	4.43	**9.23**	8.78	6.25
S_2	**414.00**	435.17	611.33	**6.54**	6.23	4.43	**9.23**	8.78	6.25
S_3	145.45	**116.54**	130.27	3.97	**4.95**	4.43	5.32	**6.64**	5.94
S_4	145.45	–	**130.27**	3.97	–	**4.43**	5.32	–	**5.94**
S_5	**414.00**	435.17	–	**6.54**	6.23	–	**9.23**	8.78	–
S_6	**414.00**	435.17	–	**6.54**	6.23	–	**9.23**	8.78	–
S_7	682.55	–	**611.33**	3.97	–	4.43	5.60	–	**6.25**
S_8	682.55	**546.86**	611.33	3.97	**4.95**	4.43	5.60	**6.99**	6.25
S_9	–	**1077.30**	–	–	**4.95**	–	–	**8.72**	–

Table A.2 PMU 16 mW switches sizing

Switch	R_{ON} (Ω)	W (μm)	C_{GG} (fF)
mS_1	414.0	6.54	9.23
S_2	414.0	6.54	9.23
S_3	116.5	4.95	6.64
S_4	130.3	4.43	5.94
S_5	414.0	6.54	9.23
S_6	414.0	6.54	9.23
S_7	611.3	4.43	6.25
S_8	546.9	4.95	6.99
S_9	1077.3	4.95	8.72

shows the required ON resistance value in each CR configuration for the switches and the corresponding transistor size and resulting gate capacitance. The lowest R_{ON} values are chosen for each CR, which are marked in bold. Table A.2 summarizes these values, which were the ones used in the prototype.

Printed in the United States
by Baker & Taylor Publisher Services